FOREST CONSERVATION AND MANAGEMENT

Edited by

Judith Rosales

ArclerPress

Forest Conservation and Management

Edited by Judith Rosales

Edition 2018

ArclerPress

Published by
Arcler Press LLC
708 3rd Avenue, 6th Floor
New York
NY 10017
United States of America
www.arclerpress.com

ISBN: 978-1-68094-624-6 (Hardback)

Printed in United States of America

Library of Congress Control Number: 2017933220

About the Editor

Judith Rosales

Prof. Judith is an Ecologist and Environmental Specialist with more than 30 years of academic and consulting experience. She holds a PhD in geography from the University of Birmingham UK and wrote the PhD and Master Academic Programs of Environmental Sciences for the Universidad de Guayana, Venezuela.

Table of Contents

3 WILDERNESS

4 DEFORESTATION

5 SLASH-AND-BURN TECHNIQUE

6 FOREST ECOLOGY

7 WILDLIFE MANAGEMENT

8 FOREST RESTORATION

9 SUSTAINABLE FOREST MANAGEMENT

List of Abbreviations

AFRSP	Ashland Forest Resiliency Stewardship Project
CCAFS	Climate Change, Agriculture and Food Security
COP	Conference of Parties
CBD	Convention on Biological Diversity
CITES	Convention on International Trade in Endangered Species of Fauna and Flora
FAO	Food and Agriculture Organization
FAO	Food and Agriculture Organization of the United Nations
FLR	Forest landscape restoration
FSC	Forest Stewardship Council
FT	Forest transition
HRD	High Rates of Deforestation
IUCN	International Union for Conservation of Nature
LFHD	Low Forest Cover with High Rates of Deforestation
NFFs	National forest funds
NCC	Natural Capital Committee
NGOs	Non-government organizations
PES	Payments for such ecological services
PEFC	Programme for the Endorsement of Forest Certification
REDD	Reducing emissions from deforestation and forest degradation
SFM	Sustainable forest management
UNEP	United Nations Environment Programme
USAID	U.S. Agency for International Development
VPD	Vapor pressure deficit

WTF/WCPA	Wilderness Task Force within the World Commission on Protected Areas
IUCN	World Conservation Union

Preface

Forest management is the maintaining and management of not only the trees in the forest, but the streams, habitat, watersheds, and even the decaying trees or logs on the forest floor. Managing our forests is not only important to the wildlife, but to our future economy and way of life. We need to continue to save the Oregon forests and help the ecosystems within them because human beings are also part of the ecosystem. By using forest management, it can help certain species of wildlife. The forest is a natural system that can supply different products and services. The working of this system is influenced by the natural environment: climate, topography, soil, etc., and also by human activity. The actions of humans in forests constitute forest management. In developed societies, this management tends to be elaborate and planned in order to achieve the objectives that are considered desirable. Sustainable forest management addresses forest degradation and deforestation while increasing direct benefits to people and the environment. At the social level, sustainable forest management contributes to livelihoods, income generation and employment. At the environmental level, it contributes to important services such as carbon sequestration and water, soil and biodiversity conservation.

Some forests have been and are managed to obtain traditional forest products such as firewood, fiber for paper, and timber, with little thinking for other products and services. Nevertheless, as a result of the progression of environmental awareness, management of forests for multiple use is becoming more common. Managing forests sustainably means increasing their benefits, including timber and food, to meet society's needs in a way that conserves and maintains forest ecosystems for the benefit of present and future generations. Many of the world's forests and woodlands, especially in the tropics and subtropics, are still not managed sustainably. Some countries lack appropriate forest policies, legislation, institutional frameworks and incentives to promote sustainable forest management, while others may have inadequate funding and lack of technical capacity. Where forest management plans exist, they are sometimes limited to ensuring the sustained production of wood, without paying attention to the many other products and services that forests offer. Forests are influenced by climate, landform and soil composition and they exist in a wide variety of forms in the tropical, temperate and boreal zones of the world. At the same time, other land uses such as agriculture can seem financially more attractive in the short term than forest management, motivating deforestation and land-use changes. Besides the need to set aside conservation areas, it is now being increasingly realised that sustainable production of wood, through environmentally sound selective harvesting practices is one of the most effective ways in ensuring in-situ conservation of the biological diversity of forest ecosystems.

Content Coverage

Chapters One and *Two* present an introduction of forest and afforestation, correspondingly. Forests are the dominant terrestrial ecosystem of Earth, and are distributed across the globe.

Chapters Three and *Four* present a coverage of wilderness and deforestation, consistently. Wilderness or wildland is a natural environment on Earth that has not been significantly modified by civilized human activity, whereas deforestation, clearance or clearing is the removal of a forest or stand of trees where the land is thereafter converted to a non-forest use

Chapters Five and *Six* are intended to focus on slash-and-burn technique and forest ecology, separately. Slash-and-burn is a forestry technique that involves cutting down trees and shrubs on a large swath of land, allowing them to dry, and then setting a fire, while forest ecology is about how organisms react with each other, and interact with the environment of a forest ecosystem.

Chapters Seven and *Eight* clarify the fundamentals of wildlife management and forest restoration, separately. Wildlife management attempts to balance the needs of wildlife with the needs of people using the best available science. Forest restoration accelerate recovery of forest structure, ecological functioning and biodiversity levels towards those typical of climax forest.

Chapter Nine provides an approach on sustainable forest management to keep the balance between three main pillars: ecological, economic and socio-cultural.

CHAPTER 1

Forests

Chapter Contents

Forest, a dense growth of trees, together with other plants, covering a large area of land. The science concerned with the study, preservation, and management of forests is forestry. A forest is an ecosystem—a community of plants and animals interacting with one another and with the physical environment (see ecology). The forests of the world are classified in three general types, or formations, which are primarily expressions of the climate in which the vegetation grows.

INTRODUCTION

A **forest** is a large area dominated by trees. Hundreds of more precise definitions of forest are used throughout the world, incorporating factors such as tree density, tree height, land use, legal standing and ecological function. According to the widely used Food and Agriculture Organization definition, forests covered four billion hectares (15 million square miles) or approximately 30 percent of the world's land area in 2006.

A conifer forest in the Swiss Alps (National Park).

The Adirondack Mountains of Upstate New York form the southernmost part of the Eastern forest-boreal transition ecoregion.

Forest on Mount Dajt, Albania.

Forests are the dominant terrestrial ecosystem of Earth, and are distributed across the globe. Forests account for 75% of the gross primary productivity of the Earth's biosphere, and contain 80% of the Earth's plant biomass.

Forests at different latitudes and elevations form distinctly different ecozones: boreal forests near the poles, tropical forests near the equator and temperate forests at mid-latitudes. Higher elevation areas tend to support forests similar to those at higher latitudes, and amount of precipitation also affects forest composition.

Human society and forests influence each other in both positive and negative ways. Forests provide ecosystem services to humans and serve as tourist attractions. Forests can also affect people's health. Human activities, including harvesting forest resources, can negatively affect forest ecosystems.

DEFINITION

Although *forest* is a term of common parlance, there is no universally recognised precise definition, with more than 800 definitions of forest used around the world. Although a forest is usually defined by the presence of trees, under many definitions an area completely lacking trees may still be considered a forest if it grew trees in the past, will grow trees in the future, or was legally designated as a forest regardless of vegetation type.

Forest in the Scottish Highlands

There are three broad categories of forest definitions in use: administrative, land use, and land cover. Administrative definitions are based primarily upon the legal designations of land, and commonly bear little relationship to the vegetation growing on the land: land that is legally designated as a forest is defined as a forest even if no trees are growing on it.Land use definitions are based upon the primary purpose that the land serves. For example, a forest may defined as any land that is used primarily for production of timber. Under such a land use definition, cleared roads or infrastructure within an area used for forestry, or areas within the region that have been cleared by harvesting, disease or fire are still considered forests even if they contain no trees. Land cover definitions define forests based upon the type and density of vegetation growing on the land. Such definitions typically define a forest as an area growing trees above some threshold. These thresholds are typically the number of trees per area (density), the area of ground under the tree canopy (canopy cover) or the section of land that is occupied by the cross-section of tree trunks (basal area). Under such land cover definitions, and area of land only be defined as forest if it is growing trees. Areas that fail to meet the land cover definition may be still included under while immature trees are establishing if they are expected to meet the definition at maturity.

Under land use definitions, there is considerable variation on where the cutoff points are between a forest, woodland, and savanna. Under some definitions, forests require very high levels of tree canopy cover, from 60% to 100%, excluding savannas and woodlands in which trees have a lower canopy cover. Other definitions consider savannas to be a type of forest, and include all areas with tree canopies over 10%.

Conversely some areas covered in trees are legally defined as agricultural areas. For example, Norway spruce plantations in Austrian forest law when the trees are being grown as Christmas trees and below a certain height.

ETYMOLOGY

The word *forest* comes from Middle English, from Old French *forest* (also *forès*) "forest, vast expanse covered by trees"; first introduced in English as the word for wild land set aside for hunting without the necessity in definition for the existence of trees. Possibly a borrowing (probably via Frankish or Old High German) of the Medieval Latin word *foresta* "open wood", *foresta* was first used by Carolingian scribes in the Capitularies of Charlemagne to refer specifically to the king's royal hunting grounds. The term was not endemic to Romance languages (e.g. native words for "forest" in the Romance languages evolved out of the Latin word *silva* "forest, wood" (English *sylvan*); cf. Italian, Spanish, Portuguese *selva*; Romanian *silvă*; Old French *selve*); and cognates in Romance languages, such as Italian *foresta*, Spanish and Portuguese *floresta*, etc. are all ultimately borrowings of the French word.

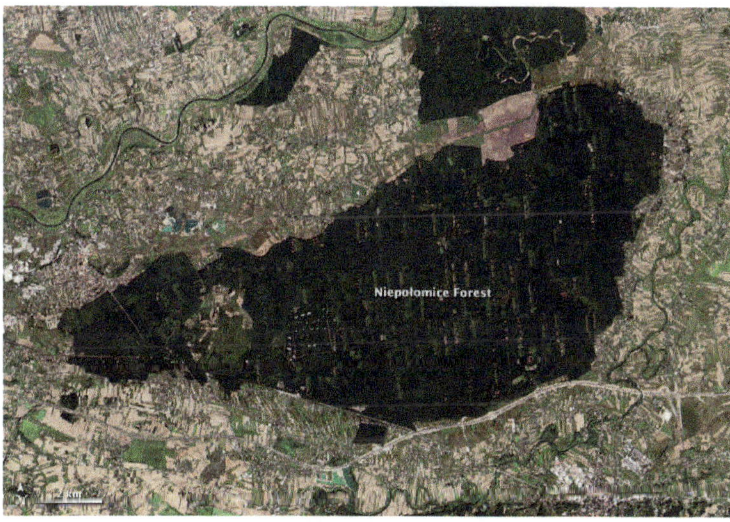

Since the 13th century, the Niepołomice Forest in Poland has had special use and protection. In this view from space, different coloration can indicate different functions.

The exact origin of Medieval Latin *foresta* is obscure. Some authorities claim the word derives from the Late Latin phrase *forestam silvam*, meaning "the outer wood"; others claim the term is a latinisation of the Frankish word **forhist* "forest, wooded country", assimilated to *forestam silvam* (a common practice among Frankish scribes). Frankish **forhist* is attested by Old High German *forst* "forest", Middle Low German *vorst* "forest", Old English *fyrhþ* "forest, woodland, game preserve, hunting ground" (English *frith*), and Old Norse *fýri* "coniferous forest", all of which derive from Proto-Germanic **furhísa-*, **furhíþija-* "a fir-wood, coniferous forest", from Proto-Indo-European **perkʷu-* "a coniferous or mountain forest, wooded height".

Uses of the word "forest" in English to denote any uninhabited area of non-enclosure are now considered archaic. The word was introduced by the Norman rulers of England as a legal term (appearing in Latin texts like the Magna Carta) denoting an uncultivated area legally set aside for hunting by feudal nobility (see Royal Forest).

These hunting forests were not necessarily wooded much, if at all. However, as hunting forests did often include considerable areas of woodland, the word "forest" eventually came to mean wooded land more generally By the start of the 14th century, the word appeared in English texts, indicating all three senses: the most common one, the legal term and the archaic usage. Other terms used to mean "an area with a high density of trees" are *wood, woodland, wold, weald, holt, frith* and *firth*. Unlike *forest*, these are all derived from Old English and were not borrowed from another language. Some classifications now reserve the term *woodland* for an area with more open space between trees and distinguish among woodlands, *open forests*, and *closed forests* based on crown cover.

Tywi forest, in Wales.

EVOLUTION

The first known forests on Earth arose in the Late Devonian (approximately 380 million years ago), with the evolution of *Archaeopteris*. *Archaeopteris* was a plant that was both tree-like and fern-like, growing to 10 metres (33 ft) in height. *Archaeopteris* quickly spread throughout the world, from the equator to subpolar latitudes. *Archaeopteris* formed the first forest by being the first known species to cast shade due to its fronds and forming soil from its roots. *Archaeopteris* was deciduous, dropping its fronds onto the forest floor. The shade, soil, and forest duff from the dropped fronds created the first forest. The shed organic matter altered the freshwater environment, slowing it down and providing food. This promoted freshwater fish.

ECOLOGY

Forests account for 75% of the gross primary productivity of the Earth's biosphere, and contain 80% of the Earth's plant biomass. Forest ecosystems can be found in all regions capable of sustaining tree growth, at altitudes up to the tree line, except where natural fire frequency or other disturbance is too high, or where the environment has been altered by human activity.

Temperate rainforest in Tasmania's Hellyer Gorge.

The latitudes 10° north and south of the equator are mostly covered in tropical rainforest, and the latitudes between 53°N and 67°N have boreal forest. As a general rule, forests dominated by angiosperms (*broadleaf forests*) are more species-rich than those dominated by gymnosperms (*conifer*, *montane*, or *needleleaf forests*), although exceptions exist.

Forests sometimes contain many tree species within a small area (as in tropical rain and temperate deciduous forests), or relatively few species over large areas (e.g., taiga and arid montane coniferous forests). Forests are often home to many animal and plant species, and biomass per unit area is high compared to other vegetation communities. Much of this biomass occurs below ground in the root systems and as partially decomposed plant detritus. The woody component of a forest contains lignin, which is relatively slow to decompose compared with other organic materials such as cellulose or carbohydrate.

Components

A forest consists of many components that can be broadly divided into two categories that are biotic (living) and abiotic (non-living) components. The living parts include trees, shrubs, vines, grasses and other herbaceous (non-woody) plants, mosses, algae, fungi, insects, mammals, birds, reptiles, amphibians, and microorganisms living on the plants and animals and in the soil.

Even, dense old-growth stand of beech trees (*Fagus sylvatica*) prepared to be regenerated by their saplings in the understory, in the Brussels part of the Sonian Forest.

Layers

A forest is made up of many layers. Starting from the ground level and moving up, the main layers of all forest types are the forest floor, the understory and the canopy. The emergent layer exists in tropical rainforests. Each layer has a different set of plants and animals depending upon the availability of sunlight, moisture and food.

Biogradska forest in Montenegro

Spiny forest at Ifaty, Madagascar, featuring various *Adansonia* (baobab) species, *Alluaudia procera* (Madagascar ocotillo) and other vegetation

- **Forest floor** contains decomposing leaves, animal droppings, and dead trees. All of decay on the forest floor and form the new soil and provides nutrients to the plants. The forest floor supports ferns, grasses, mushroom and trees seedlings.
- **Understory** is made up of bushes, shrubs, and young trees that are adapted to living in the shades of the canopy.
- **Canopy** is formed by the mass of intertwined branches, twigs and leaves of the mature trees. The crowns of the dominant trees receive most of the sunlight. This is the most productive part of the trees where maximum food is produced. The canopy forms a shady, protective "umbrella" over the rest of the forest.
- **Emergent layer** exists in the tropical rain forest and is composed of a few scattered trees that tower over the canopy.

Types

Forests can be classified in different ways and to different degrees of specificity. One such way is in terms of the biome in which they exist, combined with leaf longevity of the dominant species (whether they are evergreen or deciduous). Another distinction is whether the forests are composed predominantly of broadleaf trees, coniferous (needle-leaved) trees, or mixed.

- Boreal forests occupy the subarctic zone and are generally evergreen and coniferous.
- Temperate zones support both broadleaf deciduous forests (*e.g.*, temperate deciduous forest) and evergreen coniferous forests (*e.g.*, temperate coniferous forests and temperate rainforests). Warm temperate zones support broadleaf evergreen forests, including laurel forests.

- Tropical and subtropical forests include tropical and subtropical moist forests, tropical and subtropical dry forests, and tropical and subtropical coniferous forests.
- Physiognomy classifies forests based on their overall physical structure or developmental stage (e.g. old growth vs. second growth).
- Forests can also be classified more specifically based on the climate and the dominant tree species present, resulting in numerous different forest types (e.g., Ponderosa pine/Douglas-fir forest).

The number of trees in the world, according to a 2015 estimate, is 3 trillion, of which 1.4 trillion are in the tropics or sub-tropics, 0.6 trillion in the temperate zones, and 0.7 trillion in the coniferous boreal forests. The estimate is about eight times higher than previous estimates, and is based on tree densities measured on over 400,000 plots. It remains subject to a wide margin of error, not least because the samples are mainly from Europe and North America.

Forests can also be classified according to the amount of human alteration. Old-growth forest contains mainly natural patterns of biodiversity in established seral patterns, and they contain mainly species native to the region and habitat. In contrast, secondary forest is regrowing forest following timber harvest and may contain species originally from other regions or habitats.

Different global forest classification systems have been proposed, but none has gained universal acceptance. UNEP-WCMC's forest category classification system is a simplification of other more complex systems (e.g. UNESCO's forest and woodland 'subformations'). This system divides the world's forests into 26 major types, which reflect climatic zones as well as the principal types of trees. These 26 major types can be reclassified into 6 broader categories: temperate needleleaf; temperate broadleaf and mixed; tropical moist; tropical dry; sparse trees and parkland; and forest plantations. Each category is described as a separate section below.

Temperate Needleleaf

Temperate needleleaf forests mostly occupy the higher latitude regions of the Northern Hemisphere, as well as high altitude zones and some warm temperate areas, especially on nutrient-poor or otherwise unfavourable soils. These forests are composed entirely, or nearly so, of coniferous species (Coniferophyta). In the Northern Hemisphere pines *Pinus*, spruces *Picea*, larches *Larix*, firs *Abies*, Douglas firs *Pseudotsuga* and hemlocks *Tsuga*, make up the canopy, but other taxa are also important. In the Southern Hemisphere, most coniferous trees (members of the Araucariaceae and Podocarpaceae) occur in mixtures with broadleaf species, and are classed as broadleaf and mixed forests.

Temperate Broadleaf and Mixed

Broadleaf forest in Bhutan.

Temperate broadleaf and mixed forests include a substantial component of trees in the Anthophyta. They are generally characteristic of the warmer temperate latitudes, but extend to cool temperate ones, particularly in the southern hemisphere. They include such forest types as the mixed deciduous forests of the United States and their counterparts in China and Japan, the broadleaf evergreen rainforests of Japan, Chile and Tasmania, the sclerophyllous forests of Australia, central Chile, the Mediterranean and California, and the southern beech Nothofagus forests of Chile and New Zealand.

Tropical Moist

There are many different types of tropical moist forests, with lowland evergreen broad leaf tropical rainforests, for example várzea and igapó forests and the terra firma forests of the Amazon Basin; the peat swamp forests, dipterocarp forests of Southeast Asia; and the high forests of the Congo Basin. Seasonal tropical forests, perhaps the best description for the colloquial term "jungle", typically range from the rainforest zone 10 degrees north or south of the equator, to the Tropic of Cancer and Tropic of Capricorn. Forests located on mountains are also included in this category, divided largely into upper and lower montane formations on the basis of the variation of physiognomy corresponding to changes in altitude.

Tropical Dry

Tropical dry forests are characteristic of areas in the tropics affected by seasonal drought. The seasonality of rainfall is usually reflected in the deciduousness of the forest canopy, with most trees being leafless for several months of the year. However, under some conditions, e.g. less fertile soils or less predictable drought regimes, the proportion of evergreen species increases and the forests are characterised as "sclerophyllous". Thorn forest, a dense forest of low stature with a high frequency of thorny or spiny species, is found where drought is prolonged, and especially where grazing animals are plentiful. On very poor soils, and especially where fire or herbivory are recurrent phenomena, savannas develop.

Sparse Trees and Parkland

Sparse trees and savanna are forests with lower canopy cover of trees. They occur principally in areas of transition from forested to non-forested landscapes. The two major zones in which these ecosystems occur are in the boreal region and in the seasonally dry tropics. At high latitudes, north of the main zone of boreal forest, growing conditions are not adequate to maintain a continuous closed forest cover, so tree cover is both sparse and discontinuous. This vegetation is variously called open taiga, open lichen woodland, and forest tundra. A savanna is a mixed woodland grassland ecosystem characterized by the trees being sufficiently widely spaced so that the canopy does not close. The open canopy allows sufficient light to reach the ground to support an unbroken herbaceous layer consisting primarily of grasses. Savannas maintain an open canopy despite a high tree density.

Taiga forest near Saranpaul in the northeast Ural Mountains, Khanty–Mansia, Russia. Trees include *Picea obovata* (dominant on right bank), *Larix sibirica*, *Pinus sibirica*, and *Betula pendula*.

Forest Plantations

Forest plantations are generally intended for the production of timber and pulpwood. Commonly mono-specific and/or composed of introduced tree species, these ecosystems are not generally important as habitat for native biodiversity. However, they can be managed in ways that enhance their biodiversity protection functions and they can provide ecosystem services such as maintaining nutrient capital, protecting watersheds and soil structure, and storing carbon.

SOCIETAL SIGNIFICANCE

Forests provide a diversity of ecosystem services including converting carbon dioxide into oxygen and biomass, acting as a carbon sink, aiding in regulating climate, purifying water, mitigating natural hazards such as floods, and serving as a genetic reserve. Forests also serve as a source of lumber and as recreational areas.

Coastal Douglas fir woodland in northwest Oregon.

Redwood tree in northern California redwood forest, where many redwood trees are managed for preservation and longevity, rather than being harvested for wood production

A forest near Vinitsa, Republic of Macedonia

Some researchers state that forests do not only provide benefits, but can in certain cases also incur costs to humans.Forests may impose an economic burden, diminish the enjoyment of natural areas, reduce the food producing capacity of grazing land and cultivated land, reduce biodiversity reduce available water for humans and wildlife, harbour dangerous or destructive wildlife, and act as reservoirs of human and livestock disease.

The management of forests is often referred to as forestry. Forest management has changed considerably over the last few centuries, with rapid changes from the 1980s onwards culminating in a practice now referred to as sustainable forest management. Forest ecologists concentrate on forest patterns and processes, usually with the aim of elucidating cause-and-effect relationships. Foresters who practice sustainable forest management focus on the integration of ecological, social, and economic values, often in consultation with local communities and other stakeholders.

Humans have generally decreased the amount of forest worldwide. Anthropogenic factors that can affect forests include logging, urban sprawl, human-caused forest fires, acid rain, invasive species, and the slash and burn practices of swidden agriculture or shifting cultivation. The loss and re-growth of forest leads to a distinction between two broad types of forest, primary or old-growth forest and secondary forest. There are also many natural factors that can cause changes in forests over time including forest fires, insects, diseases, weather, competition between species, etc. In 1997, the World Resources Institute recorded that only 20% of the world's original forests remained in large intact tracts of undisturbed forest. More than 75% of these intact forests lie in three countries—the boreal forests of Russia and Canada and the rainforest of Brazil.

In 2010, the Food and Agriculture Organization of the United Nations reported that world deforestation, mainly the conversion of tropical forests to agricultural land, had decreased over the past ten years but still continues at a high rate in many countries. Globally, around 13 million hectares of forests were converted to other uses or lost through natural causes each year between 2000 and 2010 as compared to around 16 million hectares per year during the 1990s. The study covered 233 countries and areas. Brazil and Indonesia, which had the highest loss of forests in the 1990s, have significantly reduced their deforestation rates. China instituted a ban on logging, beginning in 1998, due to the erosion and flooding that it caused. In addition, ambitious tree planting programmes in countries such as China, India, the United States and Vietnam - combined with natural expansion of forests in some regions - have added more than seven million hectares of new forests annually. As a result, the net loss of forest area was reduced to 5.2 million hectares per year between 2000 and 2010, down from 8.3 million hectares annually in the 1990s. In 2015, a study for *Nature Climate Change* showed that the trend has recently been reversed, leading to an "overall gain" in global biomass and forests. This gain is due especially to reforestation in China and Russia. However new forests are not completely equivalent to old growth forests in terms of species diversity, resilience and carbon capture. On September 7, 2015, the Food and Agriculture Organization of the United Nations released a new study stating that, over the last 25 years, the global deforestation rate has decreased by 50% due to improved management of forests and greater government protection.

Smaller areas of woodland in cities may be managed as urban forestry, sometimes within public parks. These are often created for human benefits; Attention Restoration Theory argues that spending time in nature reduces stress and improves health, while forest schools and kindergartens help young people to develop social as well as scientific skills in forests. These typically need to be close to where the children live, for practical logistics.

Canada

Canada has about 4,020,000 square kilometres (1,550,000 sq mi) of forest land. More than 90% of forest land is publicly owned and about 50% of the total forest area is allocated for harvesting. These allocated areas are managed using the principles of sustainable forest management, which includes extensive consultation with local stakeholders. About eight percent of Canada's forest is legally protected from resource development. Much more forest land—about 40 percent of the total forest land base—is subject to varying degrees of protection through processes such as integrated land use planning or defined management areas such as certified forests.

By December 2006, over 1,237,000 square kilometers of forest land in Canada (about half the global total) had been certified as being sustainably managed. Clearcutting, first used in the latter half of the 20th century, is less expensive, but devastating to the environment, and companies are required by law to ensure that harvested areas are adequately regenerated. Most Canadian provinces have regulations limiting the size of clearcuts, although some older clearcuts can range upwards of 110 square kilometres (27,000 acres) in size which were cut over several years.

Garibaldi Provincial Park, British Columbia

UNITED STATES

In the United States, most forests have historically been affected by humans to some degree, though in recent years improved forestry practices has helped regulate or moderate large scale or severe impacts. However, the United States Forest Service estimates a net loss of about 2 million hectares (4,942,000 acres) between 1997 and 2020; this estimate includes conversion of forest land to other uses, including urban and suburban development, as well as afforestation and natural reversion of abandoned crop and pasture land to forest. However, in many areas of the United States, the area of forest is stable or increasing, particularly in many northern states. The opposite problem from flooding has plagued national forests, with loggers complaining that a lack of thinning and proper forest management has resulted in large forest fires.

Trees on a mountain in northern Utah during early autumn.

LAND AREA

Forest land area										
	2008		**2009**		**2010**		**2011**		**2012**	
	('000 km^2)	**('000 mi^2)**	**('000 km^2)**	**('000 mi^2)**	**('000 km^2)**	**('000 mi^2)**	**('000 km^2)**	**('000 mi^2)**	**('000 km^2)**	**('000 mi^2)**
Australia	1,511	583	1,502	580	1,493	576	1,473	569	1,467	566
Brazil	5,239	2,023	5,217	2,014	5,195	2,006	5,183	2,001	5,173	1,997
Canada	3,101	1,197	3,101	1,197	3,101	1,197	3,101	1,197	3,101	1,197
China	2,013	777	2,041	788	2,069	799	2,159	834	2,168	837
European Union	1,559	602	1,564	604	1,569	606	1,573	607	1,578	609
Germany	111	43	111	43	111	43	113	44	113	44
India	681	263	683	264	684	264	693	268	693	268
Indonesia	958	370	951	367	944	364	937	362	931	359
Japan	250	97	250	97	250	97	251	97	251	97
Russia	8,090	3,120	8,090	3,120	8,091	3,124	8,092	3,124	8,093	3,125
United States	3,033	1,171	3,036	1,172	3,040	1,170	3,049	1,177	3,051	1,178
World total	40,318	15,567	40,261	15,545	40,204	15,523	40,184	15,515	39,519	15,258

REFERENCES

1. "Forest". Dictionary.com. Retrieved 2014-11-16.
2. Schuck, Andreas; Päivinen, Risto; Hytönend, Tuomo; Pajari, Brita (2002). "Compilation of Forestry Terms and Definitions" (PDF). Joensuu, Finland: European Forest Institute. Retrieved 2014-11-16.
3. «Definitions: Indicative definitions taken from the Report of the ad hoc technical expert group on forest biological diversity". Convention on Biological Diversity. Retrieved 2014-11-16.
4. "Forest definition and extent" (PDF). United Nations Environment Programme. 2010-01-27. Retrieved 2014-11-16.
5. Comparative framework and Options for harmonization of definitions. Second Expert Meeting on Harmonizing Forest-Related Definitions. FAO. Retrieved 2014-11-26.
6. Johnson, F.X.; Pacini, H.; Smeets, E (2013). Transformations in EU biofuels markets under the Renewable Energy Directive and the implications for land use, trade and forests. CIFOR. p. 32. ISBN 9786028693813.
7. Pan, Yude; Birdsey, Richard A.; Phillips, Oliver L.; Jackson, Robert B. (2013). "The Structure, Distribution, and Biomass of the World's Forests" (PDF). Annu. Rev. Ecol. Evol. Syst. **44**: 593–62. doi:10.1146/annurev-ecolsys-110512-135914.
8. Vogt, Kristina A, ed. (2007). "Global Societies and Forest Legacies Creating Today's Forest Landscapes". Forests and Society: Sustainability and Life Cycles of Forests in Human Landscapes. CABI. pp. 30–59. ISBN 9781845930981.

9. MacDicken, Kenneth (2013-03-15). "Forest Resources Assessment Working Paper 180" (PDF). Rome: Food and Agriculture Organization of the United Nations Forestry Department. Retrieved 2014-11-16.
10. Watson, Robert T.; Verardo, David J.; Noble, Ian R.; Bolin, Bert; Ravindranath, N.H.; Dokken, David J., eds. (2000). "Land Use, Land-Use Change and Forestry". Intergovernmental Panel on Climate Change. Retrieved 2014-11-16.
11. Menzies, Nicholas; Grinspoon, Elisabeth (2007-10-22). "Facts on Forests and Forestry". ForestFacts.org, a subsidiary of GreenFacts.org. Retrieved 2014-11-16.
12. Introduction: Definition of a Forest". MuseumLink Illinois. Retrieved 2014-11-16.
13. A Polish Royal Forest". NASA Earth Observatory. November 29, 2013.
14. Broadsheet: The News Letter for Broadland Tree Wardens" (PDF). April 2011. Retrieved 2014-11-16.
15. Ritter, E (2011). Dauksta, D, ed. New Perspectives on People and Forests. World Forests. **9**. Springer. ISBN 94-007-1149-2.
16. "forest, noun". Oxford English Dictionary online edition. Oxford University Press. Retrieved 22 May 2009.
17. forest, noun". The American Heritage Dictionary of the English Language (3 ed.). Boston: Houghton Mifflin Company. 1996. ISBN 0-395-44895-6.
18. What is a Forest?". *Australian Government/Department of Agriculture, Fisheries, and Forestry/Rural Areas. 28 March 2007. Archived from the* originalon 9 October 2009. Retrieved 25 November 2010.
19. "The First Forests". Devonian Times. Retrieved 2016-05-28.
20. The Living World. D.A.V. College Managing Committee.
21. Amos, Jonathan (3 September 2015). "Earth's trees number 'three trillion'". BBC. Retrieved 3 September 2015.
22. Chazdon, Robin L. (2008). "Beyond deforestation: restoring forests and ecosystem services on degraded lands" (PDF). Science. **320** *(5882): 1458–1460. Bibcode*:2008Sci...320.1458C. doi:10.1126/science.1155365. PMID 18556551.
23. Jenkins, Martin D.; Groombridge, Brian (2002). World Atlas of Biodiversity: Earth's Living Resources in the 21st Century. World Conservation Monitoring Centre, United Nations Environment Programme. ISBN 9780520236684.
24. Chape, S; Spalding, M; Jenkins, M (2008). The world's protected areas: status, values and prospects in the 21st century. Univ de Castilla La Mancha. ISBN 0520246608.
25. Nasi, R; Wunder, S; Campos A, JJ (March 11, 2002). "Forest ecosystem services: can they pay our way out of deforestation?" (PDF). UNFF II. Costa Rica.
26. Emerton, Lucy (1999). Mount Kenya: The Economics Of Community Conservation (PDF) (Community Conservation research Working Paper). Evaluating Eden Series. University of Manchester Institute of Development Policy and Management.
27. Henneleen de Boo, Henk Lette (2002). "Economic Valuation of Forests and

Nature A support tool for effective decision-making". Theme Studies Series 6 Forests. Ede, The Netherlands: Forestry and Biodiversity Support Group, International Agricultural Centre (IAC), Wageningen National Reference Centre for Agriculture, Nature Management and Fisheries (EC-LNV).

28. Bishop, Joshua T., ed. (1999). Valuing Forests A Review of Methods and Applications in Developing Countries. London: Environmental Economics Programme, International Institute for Environment and Development (IIED).

29. Gray, Emma F.; Bond, William J. (2013). "Will woody plant encroachment impact the visitor experience and economy of conservation areas?". Koedoe. **55** *(1). ISSN 0075-6458.*

30. Scholes, R.J.; Archer, S.R. (1997). "Tree-Grass Interactions in Savannas"(PDF). Annual Review of Ecology and Systematics. **28**: 517–544. doi:10.1146/annurev.ecolsys.28.1.517.

31. Pimentel, David; Pimentel, Marcia H. (2007). Food, Energy, and Society. CRC Press.

32. Ratajczak, Zakary; Nippert, Jesse B.; Collins, Scott L. (2012). "Woody encroachment decreases diversity across North American grasslands and savannas" (PDF). Ecology. **93** *(4): 697–703. doi:10.1890/11-1199.1. PMID 22690619.*

33. Parr, Catherine L.; Gray, Emma F.; Bond, William J. (2012). "Cascading biodiversity and functional consequences of a global change–induced biome switch" (PDF). Diversity and Distributions. **18** *(5): 493–503. doi:10.1111/j.1472-4642.2012.00882.x.*

34. Wilcox, B.P.; Kreuter, U.P. (2003). Woody plant: streamflow interactions as a basis for land management decisions in drylands. Proceedings VIIth International Rangelands Congress. pp. 989–996.

35. Scott, D.F. (1999). "Managing riparian vegetation to sustain streamflow: results of paired catchment experiments in South Africa" (PDF). Canadian Journal of Forest Research. **29** *(7): 1149–115.* doi:10.1139/x99-042.

36. Davidson, A; Elliston, L, Kokic, P; Lawson, K (2005). "Native vegetation: cost of preservation in Australia" (PDF). Australian Commodities. **12** *(3): 543–548.*

37. Wilcox, B.A.; Ellis, B. "Forests and emerging infectious diseases of humans". Food and Agriculture Organization of the United Nations. Retrieved 2014-11-16.

38. Margaletic, J (2003). "Small rodents in the forest ecosystem as infectious disease reservoirs". Acta Med Croatica *(in Croatian).* **57** *(5): 421–6. PMID 15011471.*

39. World Resources Institute, 1998. The Last Frontier Forests: Ecosystems and Economies on the Edge

40. FAO. 2010. Global Forest Resources Assessment 2010. Main report. FAO Forestry Paper 163. Rome, Italy". Fao.org. Retrieved 3 July 2013.

41. Ban on Logging Saves Forests". 25 October 2001.

42. Liu, Yi Y.; van Dijk, Albert I.J.M.; de Jeu, Richard A.M.; Canadell, Josep G.; McCabe, Matthew F.; Evans, Jason P.; Wang, Guojie (30 March 2015). "Recent reversal in loss of global terrestrial biomass". Nature Climate Change. **5** *(5):*

470. Bibcode:2015NatCC...5..470L. doi:10.1038/nclimate2581.

43. World deforestation slows down as more forests are better managed". fao.org. Food and Agriculture Organization of the United Nation. Retrieved 2 October 2015.
44. MacDicken, K.; Jonsson, Ö.; Piña, L.; Maulo, S.; Adikari, Y.; Garzuglia, M.; Lindquist, E.; Reams, G.; D'Annunzio, R. "Global Forest Resources Assessment 2015" (PDF). fao.org. Food and Agriculture Organization of the United Nations. Retrieved 2 October 2015.
45. Canada". Global Forest Watch Canada. Retrieved 2014-11-28.
46. "Canada's Forests". Natural Resources Canada. 2014-10-14. Retrieved 2014-11-28.
47. Statistics". Certification status - Canada & the globe. Canadian Sustainable Forestry Certification Coalition. Retrieved 2014-11-28.
48. Wildfires Ignite Forest Management Debate". Wildrockiesalliance.org. Retrieved 3 July 2013.
49. Brock, Emily K. (2015). Money Trees: The Douglas Fir and American Forestry, 1900-1944. Oregon State University Press.
50. Forest Land Area". FAOSTAT. World Bank. 12 February 2014.

CHAPTER 2

Afforestation

Chapter Contents

- **Introduction**
- **Countries and Regions**
- **Agroforestry**

Afforestation is the planting of trees to create a forest on non-forest land. It is different from *reforestation*, which is replanting trees where a forest has been depleted. The main purposes for implementing afforestation are commercial forestry and environmental restoration or preservation.

When afforestation is used for commercial purposes, it usually involves planting and harvesting of trees as agricultural crops. Areas where this is done are known as *plantation forests*. Harvesting trees from plantation forests helps to reduce deforestation in natural forests. On the other hand, there may be environmental consequences if this tree-farming is done incorrectly. The fast-growing trees often chosen for plantation forests consume large amounts of water and may deplete the area's water resources.

INTRODUCTION

Afforestation is the establishment of a forest or stand of trees in an area where there was no previous tree cover. Reforestation is the reestablishment of forest cover, either naturally (by natural seeding, coppice, or root suckers) or artificially (by direct seeding or planting).

An afforestation project in Rand Wood, Lincolnshire, England.

Forestation is the establishment of forest growth on areas that either had forest or lacked it. Reforestation and afforestation are categories of forestation. Many governments and non-governmental organizations directly engage in programs of *afforestation* to create forests, increase carbon capture and carbon sequestration, and help to anthropogenically improve biodiversity. (In the UK, afforestation may mean converting the legal status of some land to "royal forest".) Special tools, e.g. tree planting bar, are used to make planting of trees easier and faster.

Biological Process

Gap dynamics refers to the pattern of plant growth that occurs following the creation of a forest gap, a local area of natural disturbance that results in an opening in the canopy of a forest. Gap dynamics are a typical characteristic of both temperate and tropical forests and have a wide variety of causes and effects on forest life.

Gaps are the result of natural disturbances in forests, ranging from a large branch breaking off and dropping from a tree, to a tree dying then falling over, bringing its roots to the surface of the ground, to landslides bringing down large groups of trees. Because of the range of causes, gaps, therefore, have a wide range of sizes, including small and large gaps. Regardless of size, gaps allow an increase in light as well as changes in moisture and wind levels, leading to differences in microclimate conditions compared to those from below the closed canopy, which are generally cooler and more shaded.

For gap dynamics to occur in naturally disturbed areas, either primary or secondary succession must occur. Ecological secondary succession is much more common and pertains to the process of vegetation replacement after a natural disturbance. Secondary succession results in second-growth or secondary forest, which currently covers more of the tropics than old-growth forest.

Since gaps let in more light and create diverse microclimates, they provide the ideal location and conditions for rapid plant reproduction and growth. In fact, most plant species in the tropics are dependent, at least in part, on gaps to complete their life cycles.

Disturbances

Broken trees create gaps in the central Amazon.

Gap dynamics are the result of disturbances within an ecosystem. There are both large scale and small scale disturbances, and both are influenced by duration and frequency. These all affect the resulting impact and regeneration patterns of the ecosystem.

The most common type of disturbance within a tropical ecosystem is fire. Since most nutrients in a tropical ecosystem are contained in the biomass of plants, fire is an important component of recycling these nutrients and therefore regenerating an ecosystem.

An example of a small scale disturbance is a tree falling. This can cause soil movement, which redistributes any nutrients or organisms that were attached to the tree. The tree falling also opens up the canopy for light entrance, which can support the growth of other trees and plants.

After a disturbance, there are several ways in which regeneration can occur. One way, termed the advance regeneration pathway, is when the primary understory already contains seedlings and saplings. This method is most common in the Neotropics when faced when small scale disturbances. The next pathway is from tree remains, or any growth from bases or roots, and is common in small disturbance gaps. The third route is referred to as the soil seed bank, and is the result of germination of seeds already found in the soil. The final regeneration pathway is the arrival of new seeds via animal dispersal or wind movement. The most critical components of the regeneration are seed distribution, germination, and survival.

Forest gaps and Forest Regeneration

Until recently, forest regeneration practices in North America have largely followed an agricultural model, with research concentrated on techniques for establishing and promoting early growth of planted stock after clearcutting (Cleary et al. 1978, Lavender et al. 1990, Wagner and Colombo 2001), followed by studies of growth and yield emphasizing single-species growth uninfluenced by overstorey canopy. Coates (2000) questioned this approach and proposed a shift to a more ecologically and socially based approach able to accommodate greater diversity in managed stands. Predictive models of forest regeneration and growth that take account of variable levels of canopy retention will be needed as the complexity of managed forest stands increases (Coates 2000).

Tree regeneration occurring inside canopy gaps after disturbance has been studied widely (Bazzaz and Pickett 1980, Platt and Strong 1989). Studies of gap dynamics have contributed much to an understanding of the role of small-scale disturbance in forest ecosystems, but they have been little used by foresters to predict tree responses following partial cutting (Coates and Burton 1997).

In high-latitude northern forests, position inside a gap can have a pronounced effect on resource levels (e.g., light availability) and microclimate conditions (e.g., soil temperature), especially along the north–south axis. Such variation must inevitably affect the amount and growth of regeneration; but relying solely on natural regeneration to separate the effects of gap size and position is problematic (Coates 2000). Among the many factors affecting seedling establishment following canopy disturbance are parent tree proximity and abundance, seedbed substrate, presence of seed consumers and dispersers, and climatic and microclimatic variability. Planted trees can be used to avoid many of the stochastic events surrounding natural seedling establishment.

Gradients of canopy influence can be created by partial cutting, and tree growth responses within gaps of various sizes and configurations, as well as within the adjacent forest matrix can form a basis for tree species selection. Hybrid spruce (the complex of white spruce, Sitka spruce, and occasionally Engelmann spruce) was one of several coniferous species used in a study in the Moist Cold subzone of the Interior Cedar–Hemlock zone in northwestern British Columbia. A total of 109 gaps were selected from a population of openings created by logging within each light and heavy partial cutting treatment in stands averaging 30 m in canopy height; 76 gaps were less than 1000 m^2, 33 were between 1000 m^2 and 5000 m^2. Canopy gap size was calculated as the area of an ellipse, the major axis of which was the longest line that could be run from canopy edge to canopy edge inside the gap, and the minor axis was the longest line that could be run from canopy edge perpendicular to the long line. Seedlings were planted in gaps and in the undisturbed and clearcut treatment units. There were strong and consistent trends in growth response among the seedlings as gap size increased. In all species, growth increased rapidly from small single-tree gaps to about 1000 m^2, but thereafter, there was little change up to 5000 m^2. Tree size and current growth rates for all species were highest in full open conditions. In large and medium gaps (300–1000 m^2), the largest trees of all species occurred in the middle gap position, with little difference between the sunny north and shady south positions, lodgepole pine excepted. The light advantage expected off the north end of higher-latitude gaps was not a benefit for tree growth, suggesting that below-ground effects of canopy edge trees have an important influence of seedling growth in these forests (Coates 2000).

In a study near Chapleau, Ontario (Groot 1992, Groot et al. 1997), openings were created in 40-year-old aspen and monitored to determine their influence on outplanted white spruce seedling development. Circular openings 9 m and 18 m in diameter, 9 m and 18 m wide east–west strips, and a 100 m × 150 m clearcut were planted and spot-seeded. The variation in solar radiation, air temperature, and soil temperature among the strips and plots was almost as great as the variation between the clearcut and intact forest. Solar radiation during the first growing season varied from 18% of the above-canopy values within the uncut stand to 68% values at the center of the 18 m strip. Near the edges of the strips, solar radiation was about 40% of the above-canopy along the south and 70% to 80% along the north. Stomatal conductance in white spruce seedlings declined generally from more sheltered to more exposed environments, correlating best with increased vapor pressure deficit (VPD). Without vegetation control, position in openings had little effect on the growth of planted white spruce; regrowth of lesser vegetation isolated seedlings from the microclimatic effects of overstorey treatment. Seedling diameters were independent of environment, while height growth was only slightly greater in environments having more light. With vegetation control, white spruce diameter and height were greatest in the center of the strips, even though there was less light there than along the north edge of the strips. Moisture stress may have accounted for that result.

Primary Succession

Succession is the slow rebuilding of forest gaps from natural or human disturbances. When major geological changes such as volcano eruptions or landslides occur, the current vegetation and soil may erode away leaving only rock. Primary succession occurs when pioneer species such as lichens colonize rock. As the lichens and mosses decompose, a soil substrate forms called peat. The peat, over time, will create a terrestrial ecosystem. From there on herbaceous, non-woody plants will develop and trees will follow. Major holes or gaps in the forest ecosystem will take hundreds of years to regenerate from a rock base.

Secondary Succession

Secondary succession occurs where a disturbance has taken place but soil remains and is able to support plant growth. It does not take nearly as long for plant regeneration to occur because of the soil substrate already present. Secondary succession is much more common than primary succession in the tropics.

Ecological secondary succession occurs in four distinct phases: First, rapid colonization of cleared land by species such as herbs, shrubs, and climbers as well as seedlings from pioneer tree species occurs and this can last up to three years. After that, short lived but fast growing shade intolerant species form a canopy over 10 to 30 years. Non-pioneer heliophilic (sun-loving) tree species then add to the biomass and species richness as well as shade tolerant species and this can last 75 to 150 years. Finally, shade-tolerant species regain full canopy stature indefinitely until another major disturbance occurs.

Cecropia trees are a common pioneer species found in gaps.

Secondary succession in the tropics begins with pioneer species, which are rapidly growing and include vines and shrubs. Once these species are established, large heliophilic species will develop such as heliconias. Cecropias are also a major pioneering tree in the tropics and they are adapted to grow well where forest gaps are giving way to sunlight. Shade-tolerant species that have remained low in the forest develop and become much taller. These successional phases do not have definite order or structure and because of the very high biodiversity in the tropics, there is a lot of competition for resources such as soil nutrients and sunlight.

Examples of Tree Dynamics

Due to the fact that horizontal and vertical heterogeneity of a forest is significantly increased by gaps, gaps become an obvious consideration in explaining high biodiversity. It has been proven that gaps create suitable conditions for rapid growth and reproduction. For example, non-shade tolerant plant species and many shade-tolerant plant species respond to gaps with an increase in growth, and at least a few species are dependent on gaps to succeed in their respected environments (Brokaw 1985; Hubbell and Foster 1986b; Murray 1988; Clark and Clark 1992). Gaps create diverse microclimates, affecting light, moisture, and wind conditions (Brokaw 1985). For example, exposure to edge effects increases a microclimate's light and wind intensity and decreases its moisture. A study conducted on Barro Colorado Island in Panama showed that gaps had greater seedling establishment and higher sapling densities than control areas.

Species richness was higher in gaps than in control areas, and there was more diversity in species composition among gaps. However, this study also found that there was a low recruitment rate per gap, which explains why gaps differed in species composition. With

2% to 3% for pioneer species and 3% to 6% for shade-tolerant and intermediate species. Suggesting that most species could not take advantage of gaps because they couldn't get to them through seed dispersal. With that said, the Janzen-Connell effect plays a major role in the tree species' relationship with gaps. The Janzen-Connell density dependent mortality model states that most trees die as seed or seedlings. In addition, host-specific predators or pathogens are predicted to be greatest where density is greatest, which is underneath parent tree. This corroborates with the major causes of gaps, which are the falling of trees due to mortality caused by termites or epiphyte growth. The Janzen-Connell model also states that balance between dispersal distance and mortality should cause highest recruitment to be at a certain distance away from the parent. Therefore, if these gaps are being created by the parents, the seedlings recruit away from the gap, resulting in increasing survival rates as the distance from the parent increases. This explains the low recruitment rate per gap found in the experiment conducted in Barro Colorado Island.

In corroboration, a study conducted in La Selva in Costa Rica calculated the crown illumination index for nine tree species ranging from gap specialists to emergent canopy species. Crown illumination values ranged from 1, which indicated low light, and 6, which indicated that the tree crown was completely exposed . After using a mathematical model to calculate the changes in tree diameter and changes in crown illumination with age. This model helped estimate life expectancy, time of passage to various sizes, and age patterns of mortality. The results showed what most gap dynamics studies show, pioneer species thrived in high light environments and non-pioneer species showed high mortality when young but the rate of mortality decreased as they aged. However, once trees were very large survivorship then decreased.

COUNTRIES AND REGIONS

Brazil

There is extensive and ongoing Amazon deforestation)

China

China has deforested most of its historically wooded areas. China reached the point where timber yields declined far below historic levels, due to over-harvesting of trees beyond sustainable yield. Although it has set official goals for reforestation, these goals are set over an 80-year time horizon and have not been significantly met by 2008. China is trying to correct these problems by projects as the Green Wall of China, which aims to replant a great deal of forests and halt the expansion of the Gobi desert. A law promulgated in 1981 requires that every school student over the age of 11 plants at least one tree per year. As a result, China has the highest afforestation rate of any country or region in the world, with 47,000 square kilometers of afforestation in 2008. However, the forest area per capita is still far lower than the international average. There has also been considerable criticism regarding the effectiveness of planting so many trees especially in regions where they never grew prior. Studies reveal that the water table of those areas is becoming deeper indicating significant water loss.

India

Afforestation in South India.

India has witnessed a minor increase in the percentage of the land area under forest cover from 1950 to 2006. In 1950 around 40.48 million hectares was covered by forest. In 1980 it increased to 67.47 million hectares and in 2006 it was found to be 69 million hectares. 23% of India is covered by forest. The forests of India are grouped into 5 major categories and 16 types based on biophysical criteria. 38% of forest is categorised as subtropical dry deciduous and 30% as tropical moist deciduous plus other smaller groups. It is taken care that only local species are planted in an area. Trees bearing fruits are preferred wherever possible due to their function as a food source.

Hong Kong

Since the founding of the crown colony in the 19th century, afforestation has taken place to prevent soil erosion in the catchment areas of the reservoirs that were built. During the Japanese occupation in the Second World War, the countryside was deforested as the remaining population required fuel to survive. Most of the trees were cut down and extensive reafforestation was carried out after the war. Trees that were planted are mostly non-native species, such as: Pinus massoniana, Acacia confusa (Formosan acacia),

Lophostemon confertus and the Paper Bark Tree.

Burkina Faso

Desertification is increasing along the Sahel, the strip of land between Africa's fertile tropics and the Sahara Desert. After a crippling famine in the 1970s caused by overgrazing and deforestation, a local community approach has been pioneered by Yacouba Sawadogo, a peasant farmer. By replanting trees and crops together in holes filled with compost, whole villages have been able to move back to areas considered uninhabitable.

Iran

Iran is considered a low forest cover region of the world with present cover approximating seven percent of the land area. This is a value reduced by an estimated six million hectares of virgin forest, which includes oak, almond and pistachio. Due to soil substrates, it is difficult to achieve afforestation on a large scale compared to other temperate areas endowed with more fertile and less rocky and arid soil conditions. Consequently, most of the afforestation is conducted with non-native species, leading to habitat destruction for native flora and fauna, and resulting in an accelerated loss of biodiversity.

JNF trees in the Negev Desert. Man-made dunes (here a liman) help keep in rainwater, creating an oasis.

Japan

Afforested botanical garden in Hattori Ryokuchi Park, Japan.

The Japanese temperate rainforest is well sustained and maintains a high biodiversity. One method that has been utilized in maintaining the health of forests in **Japan** has been **afforestation**. The Japanese government and private businesses have set up multiple projects to plant native tree species in open areas scattered throughout the country. This practice has resulted in shifts in forest structure and a healthy temperate rainforest that maintains a high biodiversity.

Purpose of Afforestation

The primary goal of afforestation projects in Japan is to develop the forest structure of the nation and to maintain the biodiversity found in the Japanese wilderness. The Japanese temperate rainforest is scattered throughout the Japanese archipelago and is home to many endemic species that are not naturally found anywhere else. As development of the country's caused a decline in forest cover, a reduction in biodiversity was seen in those areas. In an effort to counteract the observed decline in biodiversity, Japan began many afforestation projects. New tree stands were planted all over the archipelago and native species that inhabited the existing wild forests began to occupy the newly forested areas.

Afforestation projects in Japan first started after the rebuilding that followed World War II. In efforts to restore the country's infrastructure after the war, large areas of forest were clear-cut for timber and to create pastures to attract immigrant farmers. A new management plan for the forests of Japan was instated after many pastures were abandoned and there was a recognized massive decline of old growth and secondary forests. Forest plantations were created to increase the health of Japanese forests and to sustain the nation's timber industry. Afforestation was combined with changes in logging practices that called for reduced clear-cutting and low impact logging over a larger area.

Many private businesses in the country take part in other afforestation projects as a means of reducing the carbon emissions of the company. Carbon sequestration is a major incentive for businesses to plant seedlings and saplings that will store atmospheric carbon in their biomass as they grow. Companies like Japex and Toyota have planted and maintained tree stands beside their plants located across Japan; planting several thousand native trees in an effort to offset carbon emissions. Businesses also monitor the health of newly planted tree stands by tracking growth and surveying near-by wild forests as a comparison of how well the new stands are doing.

Historically, the forests of Japan were not extensively cut as a means of reducing the frequency of landslides and other natural disasters occurring. The root structure of the forests held the soil in place and stabilized the ground in an environment that experiences heavy rainfall and earthquake activity. In present afforestation projects, trees are used to anchor the soil and reduce the amount of soil erosion undergone by that area. A healthy tree stand also restricts the amount of soil disturbance due to rainfall by intercepting heavy precipitation in the canopy and by slowing the rate of run-off with the litter covering the soil. The role that the Japanese temperate rainforest plays in the prevention of soil erosion was recognized early on in the history of the nation and continues to be a positive reason to reforest open areas as new tree stands reduce the amount of nutrients lost in an area and allow for a more productive ecosystem.

The logging industry of Japan can be divided into pre-World War II and post-WWII. Prior to the Second World War, Japan has a small timber industry that impacted a small area as most of the forests were preserved to prevent erosion and maintain the overall quality of the land. It was during World War II that the logging industry really expanded as timber was increasingly in demand.

The increased demand of timber during the war continued on after it had ended. Japan continued to cut down large forested areas in order to rebuild. Old growth and secondary stands were increasingly fragmented as areas were clear-cut to allow for reconstruction. Between 1945 and 1965, there was rapid deforestation throughout the temperate rainforest of Japan.

During this twenty-year period, open pastures were cultivated in the Hokkaido region of northern Japan in an effort to attract more people to the area for agricultural development. Over forty-five thousand households immigrated to the Hokkaido region, but only 28.6% stayed. Most of the cultivated pasture land was abandoned and returned to the Japanese government from 1966-1977, as climate conditions in the area were not conducive to good crop yield. The shift from old growth forest to pasture left large areas of reduced soil fertility that trees were unable to recolonize. Due to a lack of a seed bank and competition with dwarf bamboo, human involvement was necessary to reforest the area. From 1978-2005 native trees with high growth rates were planted in plantations. It was mostly conifers that were planted in the area, but it has aided in the recovery of a conifer-broadleaf mixed forest.

In 1973, there was new forestry management implemented in Japan. It restricted clear-cutting and called for selective cutting over a large area. Wild forest was left surrounding areas that had been cut and newly planted. Though the plantation stage converted most of the old growth and secondary forests to conifer plantations, the new management pushed for reduced that shift.

Present day Japan continues to organize several afforestation programs throughout the nation. The government does continue to regulate forestry, but most of the afforestation projects are now put on by various private organizations. Also, most of the timber that Japan uses is imported from foreign markets; this allows for reduced logging in Japan and more time for Japanese forests to grow.

Processes Involved

There has been varying degrees of human involvement in afforested regions. For most of the areas that are reforested, there is a clearing of land that is planted with native saplings. Growth is generally monitored, but there is not much human involvement beyond checking the health of a stand.

In other regions, there is more land preparation involved prior to planting anything. Most open areas get quickly colonized by dwarf bamboo that out-competes any saplings trying to grow in the area. To give any new saplings a chance to grow in a new area people practice soil scarification. The land is plowed and soil is rotated. This process kills off any tall grasses of bamboo in the area, effectively reducing competition for new saplings. This process does subject the soil to compaction, making it harder for roots to establish; however, once the trees are established, they can regenerate and facilitate growth of other tree species.

To facilitate further growth and development in plantations, there is selective logging in order to create microhabitats for new growth using pit-and-mound topography. Trees are tipped over to create mounds that turn up more nutrients in the soil and allow for more sunlight in some patches of the understory. A healthy understory of young trees serves as a source of regeneration when an older, larger tree dies.

Tree Species

Japan afforestation projects plant only tree species native to Japan. Originally, species were chosen for their rapid growth rates and tolerance for multiple environmental conditions, which led to a large shift from old growth and secondary forests to conifer plantations. Current afforestation projects plant a more diverse number of species.

The dominant tree species planted are Japanese cedar (*Cryptomeria japonica*) and Japanese cypress (*Chamaecyparis obtusa*). Both conifer species are native to Japan and prefer to grow in deep, well-drained soils in warm, wet climates. Both species were able to successfully occupy the open areas made available by rapid deforestation. Other than being able to grow quickly, the trees were chosen because of their ornamental use in urban areas and because their wood is highly valuable timber.

Other native tree species are planted in an effort to regenerate mixed forests of conifers and broadleaf deciduous trees. Other species include: Mongolian oak (*Quercus mongolica*), lobed elm (*Ulmus laciniata*), Japanese white birch (*Betula platyphylla*), Glehn's spruce (*Picea glehnii*), and sakhalin fir (*Abies sachalinensis*).

Changes in Soil

One study examining the soil of various Japanese conifer plantations found that the plantations had a higher concentration of soil organic carbon than wild, natural forests in Japan. Soil carbon is associated with the amount of nutrients that a soil sample can hold. When areas undergo deforestation a lot of the soil organic carbon is exposed and vulnerable to degradation. This results in a loss of nutrients in deforested areas. It would be expected that plantations would have a much smaller amount of organic carbon in the soil because the soil was exposed and experienced a nutrient loss; however, it was found that conifer plantations in Japan actually had more soil carbon than wild forests in the area. It was determined that the heavy litter from newer conifer plantations resulted in an abundance of soil carbon. It was also noted that there was less soil carbon found in plantations that had undergone soil scarification prior to being planted. This was determined to be because soil scarification brought up and exposed more soil organic carbon and more was lost to degradation; however there was still a high concentration of soil carbon in those plantations as well.

Concerns

Afforestation is generally seen as a positive thing, and the re-establishment of forests in Japan has been positive overall. Large-scale afforestation is still a new concept, and there are some concerns associated with it. Afforested regions have impacts on local watersheds as well as species interactions.

One study examined the reduced water quality of watersheds downstream of afforested conifer plantations. The concern was that there was no improvement seen with public water despite improved water treatment in those areas. It was found that heavy rainfall upstream in afforested conifer plantations caused an increase in nitrogen and phosphorus running off into the streams. The undeveloped understory in those plantations means that a lot of the soil is exposed. The litter does slow down the rate of the run-off, but a large amount of nutrients still get picked up and put into the watershed. This resulted in an increase in particulate and dissolved nitrogen and phosphorus.

A large concern in Japan is root rot experienced in conifer species as a result of *Armillaria*, a group of fungus species found in Asia. Sites all over Japan were examined and it was found that every conifer species native to Japan was vulnerable to at least one species of *Armillaria*. This is a major concern because every conifer species is vulnerable, but it is especially concerning because the majority of the Japanese temperate rainforest is made up of conifer plantations consisting of monocultures of just a few species.

Another concern is due to Japan's practice of importing timber from other places. Although this has proven to be beneficial for Japanese forests, it has resulted in massive deforestation in places that do not have well-developed regulation on logging. Japan is one of the largest importers of timber, which has come at a cost of temperate and tropical forest decline in China and South America. As a result of this negativity, Japan has organized afforestation programs in other countries.

Israel

Tree-planting is an ancient Jewish tradition, mentioned in the Talmud as being more important than greeting the Messiah.With over 240 million planted trees, Israel is one of only two countries that entered the 21st century with a net gain in the number of trees, due to massive afforestation efforts. Israeli forests are the product of a major afforestation campaign by the Jewish National Fund (JNF).

Critics argue that many JNF lands inside the West Bank were illegally confiscated from Palestinian refugees, and that the JNF furthermore should not be involved with lands in the West Bank. Shaul Ephraim Cohen has claimed that trees have been planted to restrict Bedouin herding. Susan Nathan wrote that forests were planted on the site of abandoned Arab villages after the 1948 war.

Since 2009, the JNF has provided the Palestinian Authority with 3,000 tree seedlings for a forested area being developed on the edge of the new city of Rawabi, north of Ramallah.

North Africa

In North Africa, the Sahara Forest Project coupled with the Seawater greenhouse has been proposed. Some projects have also been launched in countries as Senegal to revert desertification. As of 2010, African leaders are discussing the combining of national resources to increase effectiveness. In addition, other projects as the Keita Project in Niger have been launched in the past, and have been able to locally revert damage done by desertification.

Europe

Europe has deforested the majority of its historical forests. The European Union (EU) has paid farmers for afforestation since 1990, offering grants to turn farmland back into forest and payments for the management of forest. Between 1993 and 1997, EU afforestation policies made possible the re-forestation of over 5,000 square kilometres of land. A second program, running between 2000 and 2006, afforested more than 1000 square kilometres of land (precise statistics not yet available). A third such program began in 2007. Europe's forests are growing by 0.8 million ha a year thanks to these programmes.

In Poland, the National Program of Afforestation was introduced by the government after World War II, when area of forests shrank to 20% of country's territory. Consequently, forested areas of Poland grew year by year, and on December 31, 2006, forests covered 29% of the country (see: Polish forests). It is planned that by 2050, forests will cover 33% of Poland.

According to Food and Agriculture Organization statistics, Spain had the third fastest afforestation rate in Europe in the 1990-2005 period, after Iceland and Ireland. In those years, a total of 44,360 square kilometers were afforested, and the total forest cover rose from 13,5 to 17,9 million hectares. In 1990, forests covered 26.6% of the Spanish territory. As of 2007, that figure had risen to 36.6%. Spain today has the fifth largest forest area in the European Union.

In January 2013 the UK government set a target of 12% woodland cover in England by 2060, up from the then 10%. Government-backed initiatives such as the Woodland Carbon Code are intended to support this objective by encouraging corporations and landowners to create new woodland to offset their carbon emissions.

Alpine and Subalpine regions have undergone a lot of deforestation and then forestation in the last 300 years. Out of this has emerged much practical experience. One example is the Rotten group, which is a method to bring in stable age mixed tree communities.

Australia

In Adelaide, South Australia (a city of 1.3 million), Premier Mike Rann (2002 to 2011) launched an urban forest initiative in 2003 to plant 3 million native trees and shrubs by 2014 on 300 project sites across the metro area. The projects range from large habitat restoration projects to local biodiversity projects. Thousands of Adelaide citizens have participated in community planting days. Sites include parks, reserves, transport corridors, schools, water courses and coastline. Only trees native to the local area are planted to ensure genetic integrity. Premier Rann said the project aimed to beautify and cool the city and make it more liveable; improve air and water quality and reduce Adelaide's greenhouse gas emissions by 600,000 tonnes of C02 a year. He said it was also about creating and conserving habitat for wildlife and preventing species loss.

United States

The United States is roughly one-third covered in forest and woodland. Nevertheless, areas in the US were subject to significant tree planting. In the 1800s people moving westward encountered the Great Plains – land with fertile soil, a growing population and a demand for timber but with few trees to supply it. So tree planting was encouraged along homesteads. Arbor Day was founded in 1872 by Julius Sterling Morton in Nebraska City, Nebraska. By the 1930s the Dust Bowl environmental disaster signified a reason for significant new tree cover. Public works programs under the New Deal saw the planting of 18,000 miles of windbreaks stretching from North Dakota to Texas to fight soil erosion (see Great Plains Shelterbelt).

At their summit in Copenhagen in 2009, organised by the UK based The Climate Group, leaders of subnational governments – states, regions and provinces – unanimously supported a recommendation by Premier Rann to plant one billion trees across their varied jurisdictions. The initiative was strongly supported by leaders present including Quebec Premier Jean Charest, California Governor Arnold Schwarzenegger and Scottish First Minister Alex Salmond. At a subsequent meeting in Rio de Janeiro in June 2012, The Climate Group announced that it had already received commitments by member governments to plant more than 500 million trees.

AGROFORESTRY

Parkland in Burkina Faso: sorghum grown under *Faidherbia albida* and *Borassus akeassii* near Banfora.

Agroforestry or **agro-sylviculture** is a land use management system in which trees or shrubs are grown around or among crops or pastureland. It combines shrubs and trees in agricultural and forestry technologies to create more diverse, productive, profitable, healthy, ecologically sound, and sustainable land-use systems.

As a Science

The theoretical base for agroforestry comes from ecology, via agroecology. From this perspective, agroforestry is one of the three principal land-use sciences. The other two are agriculture and forestry.

Agroforestry has a lot in common with intercropping. Both have two or more plant species (such as nitrogen-fixing plants) in close interaction, both provide multiple outputs, as a consequence, higher overall yields and, because a single application or input is shared, costs are reduced. Beyond these, there are gains specific to agroforestry.

Benefits

Agroforestry systems can be advantageous over conventional agricultural, and forest production methods. They can offer increased productivity, economic benefits, and more diversity in the ecological goods and services provided .(An example of this was seen in trying to conserve Milicia excelsa.)

Biodiversity in agroforestry systems is typically higher than in conventional agricultural systems. With two or more interacting plant species in a given land area, it creates a more complex habitat that can support a wider variety of birds, insects, and other animals. Depending upon the application, impacts of agroforestry can include:

- Reducing poverty through increased production of wood and other tree products for home consumption and sale
- Contributing to food security by restoring the soil fertility for food crops
- Cleaner water through reduced nutrient and soil runoff
- Countering global warming and the risk of hunger by increasing the number of drought-resistant trees and the subsequent production of fruits, nuts and edible oils
- Reducing deforestation and pressure on woodlands by providing farm-grown fuelwood
- Reducing or eliminating the need for toxic chemicals (insecticides, herbicides, etc.)
- Through more diverse farm outputs, improved human nutrition
- In situations where people have limited access to mainstream medicines, providing growing space for medicinal plants
- Increased crop stability
- Multifunctional site use i.e. crop production and animal grazing.
- Typically more drought resistant.
- Stabilises depleted soils from erosion
- Bioremediation

Agroforestry practices may also realize a number of other associated environmental goals, such as:

- Carbon sequestration
- Odour, dust, and noise reduction
- Green space and visual aesthetics
- Enhancement or maintenance of wildlife habitat

Adaptation to Climate Change

There is some evidence that, especially in recent years, poor smallholder farmers are turning to agroforestry as a mean to adapt to the impacts of climate change. A study from the CGIAR research program on Climate Change, Agriculture and Food Security (CCAFS) found from a survey of over 700 households in East Africa that at least 50% of those households had begun planting trees on their farms in a change from their practices 10 years ago. The trees ameliorate the effects of climate change by helping to stabilize erosion, improving water and soil quality and providing yields of fruit, tea, coffee, oil, fodder and medicinal products in addition to their usual harvest. Agroforestry was one of the most widely adopted adaptation strategies in the study, along with the use of improved crop varieties and intercropping.

Applications

Agroforestry represents a wide diversity in application and in practice. One listing includes over 50 distinct uses. The 50 or so applications can be roughly classified under a few broad headings. There are visual similarities between practices in different categories. This is expected as categorization is based around the problems addressed (countering winds, high rainfall, harmful insects, etc.) and the overall economic constraints and objectives (labor and other inputs costs, yield requirements, etc.). The categories include :

- Parklands
- Shade systems
- Crop-over-tree systems
- Alley cropping
- Strip cropping
- Fauna-based systems
- Boundary systems
- Taungyas
- Physical support systems
- Agroforests
- Wind break and shelterbelt.

Parkland

Parklands are visually defined by the presence of trees widely scattered over a large agricultural plot or pasture. The trees are usually of a single species with clear regional favorites. Among the beaks and benefits, the trees offer shade to grazing animals, protect crops against strong wind bursts, provide tree prunings for firewood, and are a roost for insect or rodent-eating birds.

There are other gains. Research with *Faidherbia albida* in Zambia showed that mature trees can sustain maize yields of 4.1 tonnes per hectare compared to 1.3 tonnes per hectare without these trees. Unlike other trees, Faidherbia sheds its nitrogen-rich leaves during the rainy crop growing season so it does not compete with the crop for light, nutrients and water. The leaves then regrow during the dry season and provide land cover and shade for crops.

Shade Systems

With shade applications, crops are purposely raised under tree canopies and within the resulting shady environment. For most uses, the understory crops are shade tolerant or the overstory trees have fairly open canopies. A conspicuous example is shade-grown coffee. This practice reduces weeding costs and improves the quality and taste of the coffee. Just because plants are grown under shade does not necessarily translate into lost or reduced yields. This is because the efficiency of photosynthesis drops off with increasing light intensity, and the rate of photosynthesis hardly increases once the light intensity is over about one tenth that of direct overhead sun. This means that plants under trees can still grow well even though they get less light. By having more than one level of vegetation, it is possible to get more photosynthesis, and overall yields, than with a single canopy layer.

Crop-over-tree Systems

Not commonly encountered, crop-over-tree systems employ woody perennials in the role of a cover crop. For this, small shrubs or trees pruned to near ground level are utilized. The purpose, as with any cover crop, is to increase in-soil nutrients and/or to reduce soil erosion.

Alley Cropping

Alley cropping corn fields between rows of walnut trees.

With alley cropping, crop strips alternate with rows of closely spaced tree or hedge species. Normally, the trees are pruned before planting the crop. The cut leafy material is spread over the crop area to provide nutrients for the crop. In addition to nutrients, the hedges serve as windbreaks and eliminate soil erosion.

Alley cropping has been shown to be advantageous in Africa, particularly in relation to improving maize yields in the sub-Saharan region. Use here relies upon the nitrogen fixing tree species *Sesbania sesban, euphorbia tricalii, Tephrosia vogelii, Gliricidia sepium* and *Faidherbia albida*. In one example, a ten-year experiment in Malawi showed that, by using the fertilizer tree Gliricidia (*Gliricidia sepium*) on land on which no mineral fertilizer was applied, maize yields averaged 3.3 tonnes per hectare as compared to one tonne per hectare in plots without fertilizer trees nor mineral fertilizers.

Strip Cropping

Strip cropping is similar to alley cropping in that trees alternate with crops. The difference is that, with alley cropping, the trees are in single row. With strip cropping, the trees or shrubs are planted in wide strip. The purpose can be, as with alley cropping, to provide nutrients, in leaf form, to the crop. With strip cropping, the trees can have a purely productive role, providing fruits, nuts, etc. while, at the same time, protecting nearby crops from soil erosion and harmful winds.

Fauna-based systems

Silvopasture over the years (Australia).

There are situations where trees benefit fauna. The most common examples are the silvopasture where cattle, goats, or sheep browse on grasses grown under trees. In hot climates, the animals are less stressed and put on weight faster when grazing in a cooler, shaded environment. Other variations have these animals directly eating the leaves of trees or shrubs.

There are similar systems for other types of fauna. Deer and hogs gain when living and feeding in a forest ecosystem, especially when the tree forage suits their dietary needs. Another variation, aquaforestry, is where trees shade fish ponds. In many cases, the fish eat the leaves or fruit from the trees.

Boundary Systems

There are a number of applications that fall under the heading of a boundary system. These include the living fences, the riparian buffer, and windbreaks.

- A living fence can be a thick hedge or fencing wire strung on living trees. In addition to restricting the movement of people and animals, living fences offer habitat to insect-eating birds and, in the case of a boundary hedge, slow soil erosion.
- Riparian buffers are strips of permanent vegetation located along or near active watercourses or in ditches where water runoff concentrates. The purpose is to keep nutrients and soil from contaminating surface water.
- Windbreaks reduce the velocity of the winds over and around crops. This increases yields through reduced drying of the crop and/or by preventing the crop from toppling in strong wind gusts.

A riparian buffer bordering a river in Iowa.

Taungya

Taungya is a vastly used system originating in Burma. In the initial stages of an orchard or tree plantation, the trees are small and widely spaced. The free space between the newly planted trees can accommodate a seasonal crop. Instead of costly weeding, the underutilized area provides an additional output and income. More complex taungyas use the between-tree space for a series of crops. The crops become more shade resistant as the tree canopies grow and the amount of sunlight reaching the ground declines. If a plantation is thinned in the latter stages, this opens further the between-tree cropping opportunities.

Physical Support Systems

In the long history of agriculture, trellises are comparatively recent. Before this, grapes and other vine crops were raised atop pruned trees. Variations of the physical support theme depend upon the type of vine. The advantages come through greater in-field biodiversity. In many cases, the control of weeds, diseases, and insect pests are primary motives.

Agroforests

These are widely found in the humid tropics and are referenced by different names (forest gardening, forest farming, tropical home gardens and, where short-statured trees or shrubs dominate, shrub gardens). Through a complex, diverse mix of trees, shrubs, vines, and seasonal crops, these systems achieve the ecological dynamics of a forest ecosystem. Because of their internal ecology, they tend to be less susceptible to harmful insects, plant diseases, drought, and wind damage.

Agroforestry similar methods were historically utilized by Native Americans. California Indians would prescribe burn oak and other habitats to maintain a 'pyrodiversity collecting model'. This method allowed for greater health of trees and the habitat in general.

Challenges

Agroforestry is relevant to almost all environments and is a potential response to common problems around the globe, and agroforestry systems can be advantageous compared to conventional agriculture or forestry. Yet agroforestry is not very widespread, at least according to current but incomplete USDA surveys as of November, 2013.

As suggested by a survey of extension programs in the United States, some obstacles (ordered most critical to least critical) to agroforestry adoption include:

- Lack of developed markets for products
- Unfamiliarity with technologies
- Lack of awareness of successful agroforestry examples
- Competition between trees, crops, and animals
- Lack of financial assistance
- Lack of apparent profit potential
- Lack of demonstration sites
- Expense of additional management
- Lack of training or expertise
- Lack of knowledge about where to market products
- Lack of technical assistance
- Cannot afford adoption or start up costs, including costs of time
- Unfamiliarity with alternative marketing approaches (e.g. web)
- Unavailability of information about agroforestry
- Apparent inconvenience
- Lack of infrastructure (e.g. buildings, equipment)
- Lack of equipment
- Insufficient land
- Lack of seed/seedling sources
- Lack of scientific research

Some solutions to these obstacles have already been suggested although many depend on particular circumstances which vary from one location to the next.

REFERENCES

1. SAFnet Dictionary | Definition For [afforestation]. Dictionaryofforestry.org. 2008-10-23. Retrieved 2012-02-17.
2. SAFnet Dictionary | Definition For [reforestation]. Dictionaryofforestry.org. 2008-08-13. Archived from the original on 2012-03-14. Retrieved 2012-02-17.
3. Silviculture Concepts and Applications R.D. Nyland 2nd Ed. 2002 pg 64
4. E. O. Wilson, 2002
5. G.A.McBeath, 2006
6. China to plant more trees in 2009_English_Xinhua. News.xinhuanet.com. 2009-01-09. Retrieved 2012-02-17.
7. 51.54 billion trees planted by ordinary Chinese in 27 years - People's Daily Online. English.people.com.cn. 2008-03-11. Retrieved 2012-02-17.
8. India: Environmental Profile. rainforests.mongabay.com. Retrieved 2015-05-04.
9. Wildash, Paula. "1080 Films - The Man Who Stopped The Desert". www.1080films.co.uk. Retrieved 2015-05-04.
10. J.A.Stanturf, 2004
11. President of German States Council of Education Ministers Plants Tree at Kennedy Memorial". JPost.com: Green Israel: People And The Environment. Jerusalem Post. The Jerusalem Post. July 29, 2009. Retrieved December 13, 2013.
12. Israel Forestry & Ecology. Jewish National Fund, East 69th Street, NY 10021 USA. Retrieved 29 October 2011.
13. JNF Tree Planting Center. Jewish National Fund, East 69th Street, NY 10021, USA. Retrieved 29 October 2011.
14. Dan Leon,The Jewish National Fund: How the Land Was 'Redeemed': The JNF's historical concept of exclusively Jewish land is wholly anachronistic"; Palestine-Israel Journal, Vol 12 No. 4 & Vol 13 No. 1, 05/06
15. Shaul Ephraim Cohen. "The Politics of Planting"; University of Chicago 1993 p.121
16. Nathan, Susan (2005). The Other Side of Israel: My Journey Across the Jewish/Arab Divide. New York: Nan A. Talese. pp. 130–131. ISBN 978-0-385-51456-9.
17. Gross, Tom (2009-12-02). "Building Peace Without Obama's Interference". Online.wsj.com. Retrieved 2013-12-21.
18. Combining of green walls. Afriqueavenir.org. Archived from the original on 2010-07-18. Retrieved 2012-08-26.
19. European Wood_Forest growth". www.europeanwood.org.cn. Retrieved 2015-05-04.
20. FAO Data. Blatantworld.com. Retrieved 2012-08-26.
21. Mongabay.com: Deforestation tables and charts for Spain. Rainforests.mongabay.com. Retrieved 2012-08-26.

22. United Nations Statistics Division - Environment Statistics. Unstats.un.org. Retrieved 2012-02-17.
23. Government Forestry and Woodlands" (PDF). Defra. Retrieved 13 June 2013.
24. http://www.waldwissen.net/waldwirtschaft/waldbau/bergwald/wsl_rottenaufforstung/index_DE
25. http://www.milliontrees.com.au
26. http://www.theclimategroup.org

CHAPTER 3

Wilderness

Chapter Contents

- **Introduction**
- **History**
- **National Parks**
- **Formal Wilderness Designation**
- **International Movement**
- **Extent**
- **Critique**

Land which is unsettled, uncultivated, and left largely alone is often referred to as "wilderness." Many people place spiritual, economic, or scientific value on these areas, and many nations have engaged in campaigns to keep part of their landscapes natural. The exact nature of the wilderness is a topic for debate, even among scientists, and this debate is often complicated by its romanticization in many cultures.

The word is derived from the Old English word for wild or savage, and it dates to around the 1200s. Many people think of wilderness as untamed, wild territory, like a primeval forest. Others see these places as a desolate wasteland, like the desert. The truth probably lies somewhere in between; wilderness is certainly far from desolate, as it hosts many plant and animal species, along with complex ecologies and interactions.

INTRODUCTION

Wilderness or wildland is a natural environment on Earth that has not been significantly modified by civilizedhuman activity. It may also be defined as: "The most intact, undisturbed wild natural areas left on our planet—those last truly wild places that humans do not control and have not developed with roads, pipelines or other industrial infrastructure."

Forrester Island Wilderness in the U.S. state of Alaska.

Some governments establish them by law or administrative acts, usually in land tracts that have not been modified by human action in great measure. The main feature of them is that human activity is restricted significantly. These actions seek not only to preserve what already exists, but also to promote and advance a natural expression and development. Wilderness areas can be found in preserves, conservation preserves, National Forests, National Parks and even in urban areas along rivers, gulches or otherwise undeveloped areas. These areas are considered important for the survival of certain species, biodiversity, ecological studies, conservation, solitude, and recreation. Wilderness is deeply valued for cultural, spiritual, moral, and aesthetic reasons. Some nature writers believe wilderness areas are vital for the human spirit and creativity. They may also preserve historic genetic traits and provide habitat for wild flora and fauna that may be difficult to recreate in zoos, arboretums or laboratories.

The word *wilderness* derives from the notion of "wildness"—in other words, that which is not controlled by humans. The mere presence or activity of people does not disqualify an area from being "wilderness." Many ecosystems that are, or have been, inhabited or influenced by activities of people may still be considered "wild." This way of looking at wilderness includes areas within which natural processes operate without human interference.

The WILD Foundation states that wilderness areas have two dimensions: they must be biologically intact and legally protected. The World Conservation Union (IUCN) classifies wilderness at two levels, Ia (Strict Nature Reserves) and Ib (Wilderness Areas). Most scientists and conservationists agree that no place on earth is completely untouched by humanity, either due to past occupation by indigenous people, or through global processes such as climate change. Activities on the margins of specific wilderness areas, such as fire suppression and the interruption of animal migration also affect the interior of wildernesses.

Especially in wealthier, industrialized nations, it has a specific legal meaning as well: as land where development is prohibited by law. Many nations have designated wilderness, including Australia, Canada, New Zealand, South Africa and the United States. Many new parks are currently being planned and legally passed by various Parliaments and Legislatures at the urging of dedicated individuals around the globe who believe that "in the end, dedicated, inspired people empowered by effective legislation will ensure that the spirit and services of wilderness will thrive and permeate our society, preserving a world that we are proud to hand over to those who come after us."

Perkuć Reserve in Puszcza Augustowska, Poland.

HISTORY

Innoko Wilderness, Alaska United States in the summer.

Ancient Times and Middle Ages

Looked at through the lens of the visual arts, nature and wildness have been important subjects in various epochs of world history. An early tradition of landscape art occurred in the Tang Dynasty (618-907). The tradition of representing nature *as it is* became one of the aims of Chinese painting and was a significant influence in Asian art. Artists in the tradition of Shan shui (lit. *mountain-water-picture*), learned to depict mountains and rivers "from the perspective of nature as a whole and on the basis of their understanding of the laws of nature... as if seen through the eyes of a bird." In the 13th century, Shih Erh Chi recommended avoiding painting "scenes lacking any places made inaccessible by nature."

For most of human history, the greater part of the Earth's terrain was wilderness, and human attention was concentrated in settled areas. The first known laws to protect parts of nature date back to the Babylonian Empire and Chinese Empire. Ashoka, the Great Mauryan King, defined the first laws in the world to protect flora and fauna in Edicts of Ashoka around 3rd Century B.C. In the Middle Ages, the Kings of England initiated one of the world's first conscious efforts to protect natural areas. They were motivated by a desire to be able to hunt wild animals in private hunting preserves rather than a desire to protect wilderness. Nevertheless, in order to have animals to hunt they would have to protect wildlife from subsistence hunting and the land from villagers gathering firewood. Similar measures were introduced in other European countries.

19th Century to Present

The idea of wilderness having intrinsic value emerged in the Western world in the 19th century. British artists John Constable and J. M. W. Turner turned their attention to capturing the beauty of the natural world in their paintings. Prior to that, paintings had been primarily of religious scenes or of human beings. William Wordsworth's poetry described the wonder of the natural world, which had formerly been viewed as a threatening place. Increasingly the valuing of nature became an aspect of Western culture.

By the mid-19th century, in Germany, "Scientific Conservation," as it was called, advocated "the efficient utilization of natural resources through the application of science and technology." Concepts of forest management based on the German approach were applied in other parts of the world, but with varying degrees of success. Over the course of the 19th century wilderness became viewed not as a place to fear but a place to enjoy and protect, hence came the conservation movement in the latter half of the 19th century. Rivers were rafted and mountains were climbed solely for the sake of recreation, not to determine their geographical context.

In 1861, following an intense lobbying by artists (painters), the French Waters and Forests Military Agency set an " artistic reserve " in Fontainebleau State Forest. With a total of 1 097 hectares, it is known to be the first World nature reserve.

Global conservation became an issue at the time of the dissolution of the British Empire in Africa in the late 1940s. The British established great wildlife preserves there. As before, this interest in conservation had an economic motive: in this case, big game hunting. Nevertheless, this led to growing recognition in the 1950s and the early 1960s of the need to protect large spaces for wildlife conservation worldwide. The World Wildlife Fund (WWF), founded in 1961, grew to be one of the largest conservation organizations in the world.

Early conservationists advocated the creation of a legal mechanism by which boundaries could be set on human activities in order to preserve natural and unique lands for the enjoyment and use of future generations. This profound shift in wilderness thought reached a pinnacle in the US with the passage of the Wilderness Act of 1964, which allowed for parts of U.S. National Forests to be designated as "wilderness preserves". Similar acts, such as the 1975 Eastern Wilderness Act, followed.

Nevertheless, initiatives for wilderness conservation continue to increase. There are a growing number of projects to protect tropical rainforests through conservation initiatives. There are also large-scale projects to conserve wilderness regions, such as Canada's Boreal Forest Conservation Framework. The Framework calls for conservation of 50 percent of the 6,000,000 square kilometres of boreal forest in Canada's north. In addition to the World Wildlife Fund, organizations such as the Wildlife Conservation Society, the WILD Foundation, The Nature Conservancy, Conservation International, The Wilderness Society (United States) and many others are active in such conservation efforts.

The 21st century has seen another slight shift in wilderness thought and theory. It is now understood that simply drawing lines around a piece of land and declaring it a wilderness does not necessarily make it a wilderness. All landscapes are intricately connected and what happens outside a wilderness certainly affects what happens inside it. For example, air pollution from Los Angeles and the California Central Valley affects Kern Canyon and Sequoia National Park. The national park has miles of "wilderness" but the air is filled with pollution from the valley. This gives rise to the paradox of what a wilderness really is; a key issue in 21st century wilderness thought.

A view of wilderness in Estonia.

NATIONAL PARKS

El Toro Wilderness within the Caribbean National Forest in Puerto Rico.

The creation of National Parks, beginning in the 19th century, preserved some especially attractive and notable areas, but the pursuits of commerce, lifestyle, and recreation combined with increases in human population have continued to result in human modification of relatively untouched areas. Such human activity often negatively impacts native flora and fauna. As such, to better protect critical habitats and preserve low-impact recreational opportunities, legal concepts of "wilderness" were established in many countries, beginning with the United States (see below).

The first National Park was Yellowstone, which was signed into law by U.S. President Ulysses S. Grant on 1 March 1872. The Act of Dedication declared Yellowstone a land "hereby reserved and withdrawn from settlement, occupancy, or sale under the laws of the United States, and dedicated and set apart as a public park or pleasuring ground for the benefit and enjoyment of the people."

The world's second national park, the Royal National Park, located just 32 km to the south of Sydney, Australia, was established in 1879.

The U.S. concept of national parks soon caught on in Canada, which created Banff National Park in 1885, at the same time as the transcontinental Canadian Pacific Railway was being built. The creation of this and other parks showed a growing appreciation of wild nature, but also an economic reality. The railways wanted to entice people to travel west. Parks such as Banff and Yellowstone gained favor as the railroads advertised travel to "the great wild spaces" of North America. When outdoorsman Teddy Roosevelt became president of the United States, he began to enlarge the U.S. National Parks system, and established the National Forest system.

By the 1920s, travel across North America by train to experience the "wilderness" (often viewing it only through windows) had become very popular. This led to the commercialization of some of Canada's National Parks with the building of great hotels such as the Banff Springs Hotel and Chateau Lake Louise.

Despite their similar name, national parks in England and Wales are quite different from national parks in many other countries. Unlike most other countries, in England and Wales, designation as a national park may include substantial settlements and human land uses which are often integral parts of the landscape, and land within a national park remains largely in private ownership. Each park is operated by its own national park authority.

Conservation and Preservation in 20th Century United States

By the later 19th century it had become clear that in many countries wild areas had either disappeared or were in danger of disappearing. This realization gave rise to the conservation movement in the United States, partly through the efforts of writers and activists such as John Burroughs, Aldo Leopold, and John Muir, and politicians such as U.S. President Teddy Roosevelt.

Cook Lake in the Bridger Wilderness, Bridger-Teton National Forest, Wyoming, U.S.

The idea of protecting nature for nature's sake began to gain more recognition in the 1930s with American writers like Aldo Leopold, calling for a "land ethic" and urging wilderness protection. It had become increasingly clear that wild spaces were disappearing rapidly and that decisive action was needed to save them. Wilderness preservation is central to deep ecology; a philosophy that believes in an inherent worth of all living beings, regardless of their instrumental utility to human needs.

Two different groups had emerged within the US environmental movement by the early 20th century: the conservationists and the preservationists. The initial consensus among conservationists was split into "utilitarian conservationists" later to be referred to as conservationists, and "aesthetic conservationists" or preservationists. The main representative for the former was Gifford Pinchot, first Chief of the United States Forest Service, and they focused on the *proper use of nature,* whereas the preservationists sought the *protection of nature from use.* Put another way, conservation sought to regulate human use while preservation sought to eliminate human impact altogether. The management of US public lands during the years 1960s and 70s reflected these dual visions, with conservationists dominating the Forest Service, and preservationists the Park Service

FORMAL WILDERNESS DESIGNATION

France

Since 1861, the French Waters and Forests Military Agency (Administration des Eaux et Forêts) put a strong protection on what was called the " artistic reserve " in Fontainebleau State Forest. With a total of 1 097 hectares, it is known to be the first World nature reserve.

Then in the 1950s, Integral Biological Reserves (Réserves Biologiques Intégrales, RBI) are dedicated to man free ecosystem evolution, on the contrary of Managed Biological reserves (Réserves Biologiques Dirigées, RBD) where a specific management is applied to conserve vulnerable species or threatened habitats.

Integral Biological Reserves occurs in French State Forests or City Forests and are therefore managed by the National Forests Office. In such reserves, all harvests coupe are forbidden excepted exotic species elimination or track safety works to avoid fallen tree risk to visitors (already existing tracks in or on the edge of the reserve).

At the end of 2014, there were 60 Integral Biological Reserves in French State Forests for a total area of 111 082 hectares and 10 in City Forests for a total of 2 835 hectares.

Greece

In Greece there are some parks called "ethniki drimoi" (εθνικοί δρυμοί, national forests) that are under protection of the Greek government. Such parks include: Olympus, Parnassos and Parnitha National Parks.

New Zealand

There are seven wilderness areas in New Zealand as defined by the National Parks Act 1980 and the Conservation Act 1987 that fall well within the IUCN definition. Wilderness areas cannot have any human intervention and can only have indigenous species re-introduced into the area if it is compatible with conservation management strategies.

In New Zealand wilderness areas are remote blocks of land that have high natural character. The Conservation Act 1987 prevents any access by vehicles and livestock, the construction of tracks and buildings, and all indigenous natural resources are protected. They are generally over 40,000 ha in size.

United States

The Great Swamp of New Jersey, donated for federal protection by concerned residents, was designated as the **first wilderness refuge in the United States** - *winter scene* photographed in March, 2008.

In the United States, a Wilderness Area is an area of federal land set aside by an act of Congress. Human activities in wilderness areas are restricted to scientific study and non-mechanized recreation; horses are permitted but mechanized vehicles and equipment, such as cars and bicycles, are not.

The United States was the first country to officially designate land as "wilderness" through the Wilderness Act of 1964. The Wilderness Act was—and is still—an important part of wilderness designation because it created the legal definition of wilderness and founded the National Wilderness Preservation System. The Wilderness Act defines wilderness as "an area where the earth and its community of life are untrammelled by man, where man himself is a visitor who does not remain."

Wilderness designation helps preserve the natural state of the land and protects flora and fauna by prohibiting development and providing for non-mechanized recreation only.

The first administratively protected wilderness area in the United States was the Gila National Forest. In 1922, Aldo Leopold, then a ranking member of the U.S. Forest Service, proposed a new management strategy for the Gila National Forest. His proposal was adopted in 1924, and 750 thousand acres of the Gila National Forest became the Gila Wilderness.

'The Great Swamp in New Jersey was the first formally designated wilderness refuge in the United States. It was declared a wildlife refuge on 3 November 1960. In 1966 it was declared a National Natural Landmark and, in 1968, it was given wilderness status. Properties in the swamp had been acquired by a small group of residents of the area, who donated the assembled properties to the federal government as a park for perpetual protection. Today the refuge amounts to 7,600 acres (31 km^2) that are within thirty miles of Manhattan.

Latir Peak Wilderness, taken from milepost 394 along US-285, ten miles north of Tres Piedras and 14 miles south of the New Mexico and Colorado border.

While wilderness designations were originally granted by an Act of Congress for Federal land that retained a "primeval character", meaning that it had not suffered from human habitation or development, the Eastern Wilderness Act of 1975 extended the protection of the NWPS to areas in the eastern States that were not initially considered for inclusion in the Wilderness Act. This act allowed lands that did not meet the constraints of size, roadlessness, or human impact to be designated as wilderness areas under the belief that they could be returned to a "primeval" state through preservation.

Approximately 107,500,000 acres (435,000 km^2) are designated as wilderness in the United States. This accounts for 4.82% of the country's total land area; however, 54% of that amount is found in Alaska (recreation and development in Alaskan wilderness is often less restrictive), while only 2.58% of the lower continental United States is designated as wilderness. Following the Omnibus Public Land Management Act of 2009 there are 756 separate wilderness designations in the United States ranging in size from Florida's Pelican Island at 5 acres (20,000 m^2) to Alaska's Wrangell-Saint Elias at 9,078,675 acres (36,740.09 km^2).

Western Australia

In Western Australia, a Wilderness Area is an area that has a wilderness quality rating of 12 or greater and meets a minimum size threshold of 8,000 hectares in temperate areas or 20,000 hectares in arid and tropical areas. A wilderness area is gazetted under section 62(1)(a) of the Conservation and Land Management Act 1984 by the Minister on any land that is vested in the Conservation Commission of Western Australia.

International

The World Conservation Union (IUCN) classifies wilderness at two levels, Ia (Strict Nature Preserves) and Ib (Wilderness areas). For the global standard of wilderness (1b) protection, governance and management, read Wilderness Protected Areas: Management Guidelines for IUCN Category 1b Protected Areas.

Forty-eight countries have wilderness areas established via legislative designation as IUCN protected area management Category 1b sites that do not overlap with any other IUCN designation. They are: Australia, Austria, Bahamas, Bangladesh, Bermuda, Bosnia and Herzegovina, Botswana, Canada, Cayman Islands, Costa Rica, Croatia, Cuba, Czech Republic, Democratic Republic of Congo, Denmark, Dominican Republic, Equatorial Guinea, Estonia, Finland, French Guyana, Greenland, Iceland, India, Indonesia, Japan, Latvia, Liechtenstein, Luxembourg, Malta, Marshall Islands, Mexico, Mongolia, Nepal, New Zealand, Norway, Northern Mariana Islands, Portugal, Seychelles, Serbia, Singapore, Slovakia, Slovenia, Spain, Sri Lanka, Sweden, Tanzania, United States of America, and Zimbabwe. At publication, there are 2,992 marine and terrestrial wilderness areas registered with the IUCN as solely Category 1b sites.

Twenty-two other countries have wilderness areas. These wilderness areas are established via administrative designation or wilderness zones within protected areas. Whereas the above listing contains countries with wilderness exclusively designated as Category 1b sites, some of the below-listed countries contain protected areas with multiple management categories including Category 1b. They are: Argentina, Bhutan, Brazil, Chile, Honduras, Germany, Italy, Kenya, Malaysia, Namibia, Nepal, Pakistan, Panama, Peru, Philippines, the Russian Federation, South Africa, Switzerland, Uganda, Ukraine, the United Kingdom of Great Britain and Northern Ireland, Venezuela, and Zambia.

INTERNATIONAL MOVEMENT

Monument Valley in Utah, United States.

At the forefront of the international wilderness movement has been The WILD Foundation, its founder Ian Player and its network of sister and partner organizations around the globe. The pioneer World Wilderness Congress in 1977 introduced the wilderness concept as an issue of international importance, and began the process of defining the term in biological and social contexts. Today, this work is continued by many international groups who still look to the World Wilderness Congress as the international venue for wilderness and to The WILD Foundation network for wilderness tools and action. The WILD Foundation also publishes the standard references for wilderness professionals and others involved in the issues: *Wilderness Management: Stewardship and Protection of Resources and Values*, the *International Journal of Wilderness*, *A Handbook on International Wilderness Law and Policy* and *Protecting Wild Nature on Native Lands* are the backbone of information and management tools for international wilderness issues.

The Wilderness Task Force within the World Commission on Protected Areas (WTF/WCPA) of the International Union for the Conservation of Nature (IUCN) plays a critical role in defining legal and management guidelines for wilderness at the international level and is also a clearing-house for information on wilderness issues. The IUCN Protected Areas Classification System defines wilderness as "A large area of unmodified or slightly modified land, and/or sea retaining its natural character and influence, without permanent or significant habitation, which is protected and managed so as to preserve its natural condition (Category 1b)." The WILD Foundation founded the WTF/WCPA in 2002 and remains co-chair.

EXTENT

The Ahklun Mountains and the Togiak Wilderness within the Togiak National Wildlife Refuge in the U.S. state of Alaska.

According to a major study, *Wilderness: Earth's Last Wild Places,* carried out by Conservation International, 46% of the world's land mass is wilderness. For purposes of this report, "wilderness" was defined as an area that "has 70% or more of its original vegetation intact, covers at least 10,000 square kilometers (3,900 sq mi) and must have fewer than five people per square kilometer." However, an IUCN/UNEP report published in 2003, found that only 10.9% of the world's land mass is currently a Category 1 Protected Area, that is, either a *strict nature reserve* (5.5%) or *protected wilderness* (5.4%). Such areas remain relatively untouched by humans. Of course, there are large tracts of lands in National Parks and other protected areas that would also qualify as wilderness. However, many protected areas have some degree of human modification or activity, so a definitive estimate of true wilderness is difficult.

The Wildlife Conservation Society generated a *human footprint* using a number of indicators, the absence of which indicate wildness: human population density, human access via roads and rivers, human infrastructure for agriculture and settlements and the presence of industrial power (lights visible from space). The society estimates that 26% of the Earth's land mass falls into the category of "Last of the wild." The wildest regions of the world include the tundra, the taiga, the Amazonian rain forest, the Tibetan Plateau, the Australian outback and deserts such as the Sahara, and the Gobi. However, from the 1970s, numerous geoglyphs have been discovered on deforested land in the Amazon rainforest, leading to claims about Pre-Columbian civilizations. The BBC's *Unnatural Histories* claimed that the Amazon rainforest, rather than being a pristine wilderness, has been shaped by man for at least 11,000 years through practices such as forest gardening and terra preta.

It should be noted that the percentage of land area designated "wilderness" does not necessarily reflect a measure of its biodiversity. Of the last natural wilderness areas, the taiga—which is mostly wilderness—represents 11% of the total land mass in the Northern Hemisphere. Tropical rainforest represent a further 7% of the world's land base. Estimates of the Earth's remaining wilderness underscore the rate at which these lands are being developed, with dramatic declines in biodiversity as a consequence.

CRITIQUE

Cedar Mountain Wilderness in northern Utah, United States.

The American concept of wilderness has been criticized by some nature writers. For example, William Cronon writes that what he calls a wilderness ethic or cult may "teach us to be dismissive or even contemptuous of such humble places and experiences", and that "wilderness tends to privilege some parts of nature at the expense of others", using as an example "the mighty canyon more inspiring than the humble marsh." This is most clearly visible with the fact that nearly all U.S. National Parks preserve spectacular canyons and mountains, and it was not until the 1940s that a swamp became a national park—the Everglades. In the mid-20th century national parks started to protect biodiversity, not simply attractive scenery.

Cronon also believes the passion to save wilderness "poses a serious threat to responsible environmentalism" and writes that it allows people to "give ourselves permission to evade responsibility for the lives we actually lead....to the extent that we live in an urban-industrial civilization but at the same time pretend to ourselves that our real home is in the wilderness".

Michael Pollan has argued that the wilderness ethic leads people to dismiss areas whose wildness is less than absolute. In his book *Second Nature*, Pollan writes that "once a landscape is no longer 'virgin' it is typically written off as fallen, lost to nature, irredeemable." Another challenge to the conventional notion of wilderness comes from Robert Winkler in his book, *Going Wild: Adventures with Birds in the Suburban Wilderness*. "On walks in the unpeopled parts of the suburbs," Winkler writes, "I've witnessed the same wild creatures, struggles for survival, and natural beauty that we associate with true wilderness." Attempts have been made, as in the Pennsylvania Scenic Rivers Act, to distinguish "wild" from various levels of human influence: in the Act, "wild rivers" are "not impounded", "usually not accessible except by trail", and their watersheds and shorelines are "essentially primitive".

Another source of criticism is that the criteria for wilderness designation is vague and open to interpretation. For example, the Wilderness Act states that wilderness must be roadless. The definition given for roadless is "the absences of roads which have been improved and maintained by mechanical means to insure relatively regular and continuous use." However, there have been added sub-definitions that have, in essence, made this standard unclear and open to interpretation.

Coming from a different direction, some criticism from the Deep Ecology movement argues against conflating "wilderness" with "wilderness reservations", viewing the latter term as an oxymoron that, by allowing the law as a human construct to define nature, unavoidably voids the very freedom and independence of human control that defines wilderness. True wilderness requires the ability of life to undergo speciation with as little interference from humanity as possible.

REFERENCES

1. What is a Wilderness Area". The WILD Foundation. Retrieved 2009-02-20.
2. No Man's Garden by Daniel B. Botkin p155-157
3. Wilderness Areas are Biologically Intact
4. Wilderness as a Protected Area Classification
5. WHAT IS WILDLAND? - a review". Retrieved 2009-07-14.
6. Vance G. Martin and Ian C. Player, Forward, A Handbook on International Wilderness Law and Policy
7. Chinese brush painting Asia-art.net Retrieved on: 20 May 2006.
8. History of Conservation BC Spaces for Nature. Retrieved: 20 May 2006.
9. Akamani, K. (nd). "The Wilderness Idea: A Critical Review." A Better Earth. org. Retrieved: 1 June 2006
10. Canadian Boreal Initiative Boreal Forest Conservation Framework. www. borealcanada.ca Retrieved on: 1 December 2007.
11. Mangan, E. Yellowstone, the First National Park Library of Congress, Mapping the National Parks. Retrieved on: 2010-08-12.
12. «The Magic of Yellowstone". Congressional Acts Pertaining to Yellowstone. Retrieved 7 December 2011.
13. New South Wales National Parks & Wildlife Service, «Parks & Reserves:

Royal National Park". Accessed 6 December 2011.

14. John Barry; E. Gene Frankland (2002). International Encyclopedia of Environmental Politics. Routledge. p. 161.
15. Young, Raymond A. (1982). Introduction to Forest Science. John Wiley & Sons. pp. 20–21. ISBN 0-471-06438--6.
16. 1995 & 1998 National Forests Office internal instructions in application of the last paragraph of article L. 212-2 of the French Forest Act
17. http://www.onf.fr/onf/sommaire/developpement_durable/actions/20080707-150833-949463/@@index.html
18. New Zealand Government (1987). "Conservation Act 1987 Part 4, Section 20". New Zealand Government. Retrieved 2008-10-02.
19. Malloy, Les. "Specially protected areas". Te Ara - the Encyclopedia of New Zealand. Retrieved 2008-10-02.
20. «Wilderness Act of 1964". Wilderness.net. Retrieved 7 December 2011.
21. «Aldo Leopold". Prominent Figures in Wilderness History. Retrieved 7 December 2011.
22. U.S. Fish and Wildlife Service. Great Swamp National Wildlife Refuge. Retrieved on: 7 June 2008.
23. "The Wilderness Act of 1964". Western North Carolina's Mountain Treasures. Retrieved on 16 June 2010.
24. Department of Conservation and Land Management Policy Statement No 62, Identification and management of Wilderness and surrounding areas.
25. «The World Database on Protected Areas (WDPA)" Check |url= value (help). IUCN and UNEP-WCMC. 2016.
26. Conservation International (2002) Global Analysis Finds Nearly Half The Earth Is Still Wilderness. Retrieved on 28 July 2007.
27. Chape, S., S. Blyth, L. Fish, P. Fox and M. Spalding (compilers) (2003). 2003 United Nations List of Protected Areas. IUCN, Gland, Switzerland and Cambridge, UK and UNEP-WCMC, Cambridge, UK. PDF
28. Wildlife Conservation Society. 2005. State of the Wild 2006: A Global Portrait of Wildlife, Wildlands and Oceans. Washington, D.C. Island Press. pp. 16 &17.
29. Simon Romero (14 January 2012). "Once Hidden by Forest, Carvings in Land Attest to Amazon's Lost World". The New York Times.
30. Martti Pärssinen; Denise Schaan & Alceu Ranzi (2009). "Pre-Columbian geometric earthworks in the upper Purús: a complex society in western Amazonia". Antiquity. 83 *(322): 1084–1095.*
31. «Unnatural Histories - Amazon". BBC Four.
32. University of Manitoba Taiga Biological Station. 2004. Frequently answered questions. Retrieved: 2006-07-04.

33. Rainforest Foundation US. 2006. Commonly asked questions. Retrieved: 2006-07-04.
34. The Trouble with Wilderness University of Wisconsin-Madison. Retrieved: 28 January 2007.
35. Pollan, Michael (2003). Second Nature: A Gardener's Education, p. 188. Grove Press. ISBN 978-0-8021-4011-1.
36. Winkler, Robert. (2003). Going Wild: Adventures with Birds in the Suburban Wilderness. National Geographic ISBN 978-0-7922-6168-1.
37. Pennsylvania Scenic Rivers Act (P.L. 1277, Act No. 283 as amended by Act 110, May 7, 1982)
38. Durrant, Jeffrey O. (2007). Struggle over Utah's San Rafael Swell: Wilderness, National Conservation Areas, and National Monuments. University of Arizona Press. p. 43. ISBN 0-8165-2669-9.
39. Thomas Birch (1995). George Sessions, ed. Deep Ecology for the 21st Century. Boston: Shambhala. pp. 345, 339–355. ISBN 1-57062-049-0.
40. George Sessions (1995). Deep Ecology for the 21st Century. Boston: Shambhala. pp. 323, 323–330. ISBN 1-57062-049-0.

CHAPTER 4

Deforestation

Chapter Contents

- **Introduction**
- **Causes**
- **Problems with Deforestation**
- **Economic Impact**
- **Forest Transition Theory**
- **Industrial Era**
- **Control**
- **Military Contex**
- **Public Health Context**
- **Deforestation and Climate Change**

Some of the major effects of deforestation include biodiversity loss, local climate change and global warming. Farmers often have trouble raising crops, which typically results in more problems, such as an increase in food prices. Soil erosion and cultural displacement also can occur. Those who want to address the problem have a variety of options, however.

INTRODUCTION

Deforestation, clearance or clearing is the removal of a forest or stand of trees where the land is thereafter converted to a non-forest use. Examples of deforestation include conversion of forestland to farms, ranches, or urban use. The most concentrated deforestation occurs in tropical rainforests. About 30% of Earth's land surface is covered by forests.

Satellite photograph of deforestation in progress in eastern Bolivia. Worldwide, 10 percent of wilderness areas were lost between 1990 and 2015.

In temperate mesic climates, natural regeneration of forest stands often will not occur in the absence of disturbance, whether natural or anthropogenic. Furthermore, biodiversity after regeneration harvest often mimics that found after natural disturbance, including biodiversity loss after naturally occurring rainforest destruction.

Deforestation occurs for multiple reasons: trees are cut down to be used or sold as fuel (sometimes in the form of charcoal) or timber, while cleared land is used as pasture for livestock and plantation. The removal of trees without sufficient reforestation has resulted in damage to habitat, biodiversity loss and aridity. It has adverse impacts on biosequestration of atmospheric carbon dioxide. Deforestation has also been used in war to deprive the enemy of cover for its forces and also vital resources. Modern examples of this were the use of Agent Orange by the British military in Malaya during the Malayan Emergency and the United States military in Vietnam during the Vietnam War. As of 2005, net deforestation rates have ceased to increase in countries with a per capita GDP of at least US$4,600. Deforested regions typically incur significant adverse soil erosion and frequently degrade into wasteland.

Disregard of ascribed value, lax forest management and deficient environmental laws are some of the factors that allow deforestation to occur on a large scale. In many countries, deforestation, both naturally occurring and human-induced, is an ongoing issue. Deforestation causes extinction, changes to climatic conditions, desertification, and displacement of populations as observed by current conditions and in the past through the fossil record. More than half of all plant and land animal species in the world live in tropical forests.

Between 2000 and 2012, 2.3 million square kilometres (890,000 square miles) of forests around the world were cut down. As a result of deforestation, only 6.2 million square kilometres (2.4 million square miles) remain of the original 16 million square kilometres (6 million square miles) of forest that formerly covered the Earth.

CAUSES

According to the United Nations Framework Convention on Climate Change (UNFCCC) secretariat, the overwhelming direct cause of deforestation is agriculture. Subsistence farming is responsible for 48% of deforestation; commercial agriculture is responsible for 32%; logging is responsible for 14%, and fuel wood removals make up 5%.

Experts do not agree on whether industrial logging is an important contributor to global deforestation. Some argue that poor people are more likely to clear forest because they have no alternatives, others that the poor lack the ability to pay for the materials and labour needed to clear forest. One study found that population increases due to high fertility rates were a primary driver of tropical deforestation in only 8% of cases.

Other causes of contemporary deforestation may include corruption of government institutions, the inequitable distribution of wealth and power, population growth and overpopulation, and urbanization. Globalization is often viewed as another root cause of deforestation, though there are cases in which the impacts of globalization (new flows of labor, capital, commodities, and ideas) have promoted localized forest recovery.

The last batch of sawnwood from the peat forest in Indragiri Hulu, Sumatra, Indonesia. Deforestation for oil palm plantation.

In 2000 the United Nations Food and Agriculture Organization (FAO) found that "the role of population dynamics in a local setting may vary from decisive to negligible," and that deforestation can result from "a combination of population pressure and stagnating economic, social and technological conditions."

The degradation of forest ecosystems has also been traced to economic incentives that make forest conversion appear more profitable than forest conservation. Many important forest functions have no markets, and hence, no economic value that is readily apparent to the forests' owners or the communities that rely on forests for their well-being. From the perspective of the developing world, the benefits of forest as carbon sinks or biodiversity reserves go primarily to richer developed nations and there is insufficient compensation for these services. Developing countries feel that some countries in the developed world, such as the United States of America, cut down their forests centuries ago and benefited economically from this deforestation, and that it is hypocritical to deny developing countries the same opportunities, i.e. that the poor shouldn't have to bear the cost of preservation when the rich created the problem.

Some commentators have noted a shift in the drivers of deforestation over the past 30 years. Whereas deforestation was primarily driven by subsistence activities and government-sponsored development projects like transmigration in countries like Indonesia and colonization in Latin America, India, Java, and so on, during the late 19th century and the earlier half of the 20th century, by the 1990s the majority of deforestation was caused by industrial factors, including extractive industries, large-scale cattle ranching, and extensive agriculture.

PROBLEMS WITH DEFORESTATION

Atmospheric

Illegal slash and burn practice in Madagascar, 2010.

Deforestation is ongoing and is shaping climate and geography.

Deforestation is a contributor to global warming, and is often cited as one of the major causes of the enhanced greenhouse effect. Tropical deforestation is responsible for approximately 20% of world greenhouse gas emissions.According to the Intergovernmental Panel on Climate Change deforestation, mainly in tropical areas, could account for up to one-third of total anthropogenic carbon dioxide emissions. But recent calculations suggest that carbon dioxide emissions from deforestation and forest degradation (excluding peatland emissions) contribute about 12% of total anthropogenic carbon dioxide emissions with a range from 6 to 17%. Deforestation causes carbon dioxide to linger in the atmosphere. As carbon dioxide accrues, it produces a layer in the atmosphere that traps radiation from the sun. The radiation converts to heat which causes global warming, which is better known as the greenhouse effect. Plants remove carbon in the form of carbon dioxide from the atmosphere during the process of photosynthesis, but release some carbon dioxide back into the atmosphere during normal respiration. Only when actively growing can a tree or forest remove carbon, by storing it in plant tissues. Both the decay and burning of wood releases much of this stored carbon back to the atmosphere. In order for forests to take up carbon, there must be a net accumulation of wood. One way is for the wood to be harvested and turned into long-lived products, with new young trees replacing them. Deforestation may also cause carbon stores held in soil to be released. Forests can be either sinks or sources depending upon environmental circumstances. Mature forests alternate between being net sinks and net sources of carbon dioxide (see carbon dioxide sink and carbon cycle).

In deforested areas, the land heats up faster and reaches a higher temperature, leading to localized upward motions that enhance the formation of clouds and ultimately produce more rainfall. However, according to the Geophysical Fluid Dynamics Laboratory, the models used to investigate remote responses to tropical deforestation showed a broad but mild temperature increase all through the tropical atmosphere. The model predicted <0.2 °C warming for upper air at 700 mb and 500 mb. However, the model shows no significant changes in other areas besides the Tropics. Though the model showed no significant changes to the climate in areas other than the Tropics, this may not be the case since the model has possible errors and the results are never absolutely definite.

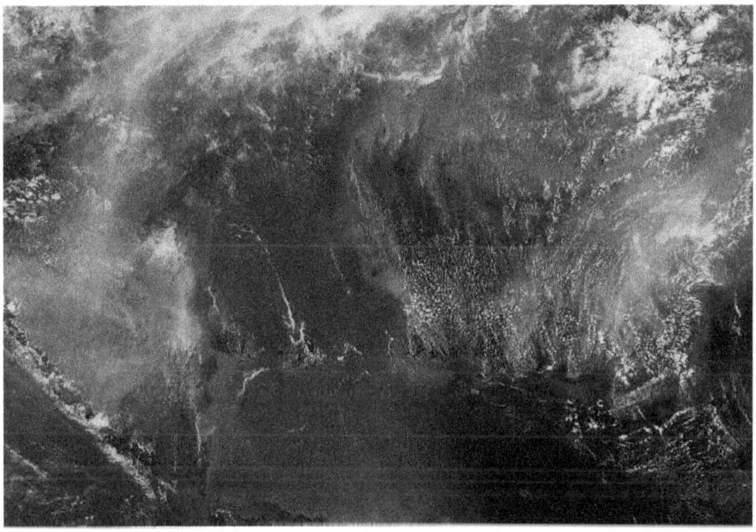

Fires on Borneo and Sumatra, 2006. People use slash-and-burn deforestation to clear land for agriculture.

Reducing emissions from deforestation and forest degradation (REDD) in developing countries has emerged as a new potential to complement ongoing climate policies. The idea consists in providing financial compensations for the reduction of greenhouse gas (GHG) emissions from deforestation and forest degradation".

Rainforests are widely believed by laymen to contribute a significant amount of the world's oxygen, although it is now accepted by scientists that rainforests contribute little net oxygen to the atmosphere and deforestation has only a minor effect on atmospheric oxygen levels. However, the incineration and burning of forest plants to clear land releases large amounts of CO_2, which contributes to global warming. Scientists also state that tropical deforestation releases 1.5 billion tons of carbon each year into the atmosphere.

Hydrological

The water cycle is also affected by deforestation. Trees extract groundwater through their roots and release it into the atmosphere. When part of a forest is removed, the trees no longer transpire this water, resulting in a much drier climate. Deforestation reduces the content of water in the soil and groundwater as well as atmospheric moisture. The dry soil leads to lower water intake for the trees to extract. Deforestation reduces soil cohesion, so that erosion, flooding and landslides ensue.

Shrinking forest cover lessens the landscape's capacity to intercept, retain and transpire precipitation. Instead of trapping precipitation, which then percolates to groundwater systems, deforested areas become sources of surface water runoff, which moves much faster than subsurface flows. Forests return most of the water that falls as precipitation to the atmosphere by transpiration. In contrast, when an area is deforested, almost all precipitation is lost as run-off. That quicker transport of surface water can translate into flash flooding and more localized floods than would occur with the forest cover. Deforestation also contributes to decreased evapotranspiration, which lessens atmospheric moisture which in some cases affects precipitation levels downwind from the deforested area, as water is not recycled to downwind forests, but is lost in runoff and returns directly to the oceans. According to one study, in deforested north and northwest China, the average annual precipitation decreased by one third between the 1950s and the 1980s.

Trees, and plants in general, affect the water cycle significantly:

- their canopies intercept a proportion of precipitation, which is then evaporated back to the atmosphere (canopy interception);
- their litter, stems and trunks slow down surface runoff;
- their roots create macropores – large conduits – in the soil that increase infiltration of water;
- they contribute to terrestrial evaporation and reduce soil moisture via transpiration;
- their litter and other organic residue change soil properties that affect the capacity of soil to store water.
- their leaves control the humidity of the atmosphere by transpiring. 99% of the water absorbed by the roots moves up to the leaves and is transpired.

As a result, the presence or absence of trees can change the quantity of water on the surface, in the soil or groundwater, or in the atmosphere. This in turn changes erosion rates and the availability of water for either ecosystem functions or human services.

The forest may have little impact on flooding in the case of large rainfall events, which overwhelm the storage capacity of forest soil if the soils are at or close to saturation.

Tropical rainforests produce about 30% of our planet's fresh water.

Soil

Undisturbed forests have a very low rate of soil loss (erosion), approximately 2 metric tons per square kilometer (6 short tons per square mile). Deforestation generally increases rates of soil loss, by increasing the amount of runoff and reducing the protection of the soil from tree litter. This can be an advantage in excessively leached tropical rain forest soils. Forestry operations themselves also increase erosion through the development of (forest) roads and the use of mechanized equipment.

China's Loess Plateau was cleared of forest millennia ago. Since then it has been eroding, creating dramatic incised valleys, and providing the sediment that gives the Yellow River its yellow color and that causes the flooding of the river in the lower reaches (hence the river's nickname 'China's sorrow').

Deforestation for the use of clay in the Brazilian city of Rio de Janeiro. The hill depicted is Morro da Covanca, in Jacarepaguá.

Removal of trees does not always increase erosion rates. In certain regions of southwest US, shrubs and trees have been encroaching on grassland. The trees themselves enhance the loss of grass between tree canopies. The bare intercanopy areas become highly erodible. The US Forest Service, in Bandelier National Monument for example, is studying how to restore the former ecosystem, and reduce erosion, by removing the trees.

Tree roots bind soil together, and if the soil is sufficiently shallow they act to keep the soil in place by also binding with underlying bedrock. Tree removal on steep slopes with shallow soil thus increases the risk of landslides, which can threaten people living nearby.

Biodiversity

Deforestation on a human scale results in decline in biodiversity, and on a natural global scale is known to cause the extinction of many species.{ The removal or destruction of areas of forest cover has resulted in a degraded environment with reduced biodiversity. Forests support biodiversity, providing habitat for wildlife; moreover, forests foster medicinal conservation. With forest biotopes being irreplaceable source of new drugs (such as taxol), deforestation can destroy genetic variations (such as crop resistance) irretrievably.

Illegal logging in Madagascar. In 2009, the vast majority of the illegally obtained rosewood was exported to China.

Since the tropical rainforests are the most diverse ecosystems on Earth and about 80% of the world's known biodiversity could be found in tropical rainforests, removal or destruction of significant areas of forest cover has resulted in a degraded environment with reduced biodiversity. A study in Rondônia, Brazil, has shown that deforestation also removes the microbial community which is involved in the recycling of nutrients, the production of clean water and the removal of pollutants.

It has been estimated that we are losing 137 plant, animal and insect species every single day due to rainforest deforestation, which equates to 50,000 species a year. Others state that tropical rainforest deforestation is contributing to the ongoing Holocene mass extinction. The known extinction rates from deforestation rates are very low, approximately 1 species per year from mammals and birds which extrapolates to approximately 23,000 species per year for all species. Predictions have been made that more than 40% of the animal and plant species in Southeast Asia could be wiped out in the 21st century. Such predictions were called into question by 1995 data that show that within regions of Southeast Asia much of the original forest has been converted to monospecific plantations, but that potentially endangered species are few and tree flora remains widespread and stable.

Scientific understanding of the process of extinction is insufficient to accurately make predictions about the impact of deforestation on biodiversity. Most predictions of forestry related biodiversity loss are based on species-area models, with an underlying assumption that as the forest declines species diversity will decline similarly. However, many such models have been proven to be wrong and loss of habitat does not necessarily lead to large scale loss of species.Species-area models are known to overpredict the number of species known to be threatened in areas where actual deforestation is ongoing, and greatly overpredict the number of threatened species that are widespread.

A recent study of the Brazilian Amazon predicts that despite a lack of extinctions thus far, up to 90 percent of predicted extinctions will finally occur in the next 40 years.

ECONOMIC IMPACT

Damage to forests and other aspects of nature could halve living standards for the world's poor and reduce global GDP by about 7% by 2050, a report concluded at the Convention on Biological Diversity (CBD) meeting in Bonn in 2008. Historically, utilization of forest products, including timber and fuel wood, has played a key role in human societies, comparable to the roles of water and cultivable land. Today, developed countries continue to utilize timber for building houses, and wood pulp for paper. In developing countries almost three billion people rely on wood for heating and cooking.

The forest products industry is a large part of the economy in both developed and developing countries. Short-term economic gains made by conversion of forest to agriculture, or over-exploitation of wood products, typically leads to loss of long-term income and long-term biological productivity. West Africa, Madagascar, Southeast Asia and many other regions have experienced lower revenue because of declining timber harvests. Illegal logging causes billions of dollars of losses to national economies annually.

The new procedures to get amounts of wood are causing more harm to the economy and overpower the amount of money spent by people employed in logging.According to a study, in most areas studied, the various ventures that prompted deforestation rarely generated more than US$5 for every ton of carbon they released and frequently returned far less than US$1. The price on the European market for an offset tied to a one-ton reduction in carbon is 23 euro (about US$35).

Rapidly growing economies also have an effect on deforestation. Most pressure will come from the world's developing countries, which have the fastest-growing populations and most rapid economic (industrial) growth. In 1995, economic growth in developing countries reached nearly 6%, compared with the 2% growth rate for developed countries." As our human population grows, new homes, communities, and expansions of cities will occur. Connecting all of the new expansions will be roads, a very important part in our daily life. Rural roads promote economic development but also facilitate deforestation. About 90% of the deforestation has occurred within 100 km of roads in most parts of the Amazon.

The European Union is one of the largest importer of products made from illegal deforestation.

FOREST TRANSITION THEORY

The forest transition and historical baselines.

The forest area change may follow a pattern suggested by the forest transition (FT) theory, whereby at early stages in its development a country is characterized by high forest cover and low deforestation rates (HFLD countries).

Then deforestation rates accelerate (HFHD, high forest cover – high deforestation rate), and forest cover is reduced (LFHD, low forest cover – high deforestation rate), before the deforestation rate slows (LFLD, low forest cover – low deforestation rate), after which forest cover stabilizes and eventually starts recovering. FT is not a "law of nature," and the pattern is influenced by national context (for example, human population density, stage of development, structure of the economy), global economic forces, and government policies. A country may reach very low levels of forest cover before it stabilizes, or it might through good policies be able to "bridge" the forest transition.

FT depicts a broad trend, and an extrapolation of historical rates therefore tends to underestimate future BAU deforestation for counties at the early stages in the transition (HFLD), while it tends to overestimate BAU deforestation for countries at the later stages (LFHD and LFLD).

Countries with high forest cover can be expected to be at early stages of the FT. GDP per capita captures the stage in a country's economic development, which is linked to the pattern of natural resource use, including forests. The choice of forest cover and GDP per capita also fits well with the two key scenarios in the FT:

(i) a forest scarcity path, where forest scarcity triggers forces (for example, higher prices of forest products) that lead to forest cover stabilization; and

(ii) an economic development path, where new and better off-farm employment opportunities associated with economic growth (= increasing GDP per capita) reduce profitability of frontier agriculture and slows deforestation.

Historical causes

Prehistory

The Carboniferous Rainforest Collapse was an event that occurred 300 million years ago. Climate change devastated tropical rainforests causing the extinction of many plant and animal species. The change was abrupt, specifically, at this time climate became cooler and drier, conditions that are not favourable to the growth of rainforests and much of the biodiversity within them. Rainforests were fragmented forming shrinking 'islands' further and further apart. This sudden collapse affected several large groups, effects on amphibians were particularly devastating, while reptiles fared better, being ecologically adapted to the drier conditions that followed.

An array of Neolithic artifacts, including bracelets, axe heads, chisels, and polishing tools.

Rainforests once covered 14% of the earth's land surface; now they cover a mere 6% and experts estimate that the last remaining rainforests could be consumed in less than 40 years. Small scale deforestation was practiced by some societies for tens of thousands of years before the beginnings of civilization. The first evidence of deforestation appears in the Mesolithic period. It was probably used to convert closed forests into more open ecosystems favourable to game animals. With the advent of agriculture, larger areas began to be deforested, and fire became the prime tool to clear land for crops. In Europe there is little solid evidence before 7000 BC. Mesolithic foragers used fire to create openings for red deer and wild boar. In Great Britain, shade-tolerant species such as oak and ash are replaced in the pollen record by hazels, brambles, grasses and nettles. Removal of the forests led to decreased transpiration, resulting in the formation of upland peat bogs. Widespread decrease in elm pollen across Europe between 8400–8300 BC and 7200–7000 BC, starting in southern Europe and gradually moving north to Great Britain, may represent land clearing by fire at the onset of Neolithic agriculture.

The Neolithic period saw extensive deforestation for farming land. Stone axes were being made from about 3000 BC not just from flint, but from a wide variety of hard rocks from across Britain and North America as well. They include the noted Langdale axe industry in the English Lake District, quarries developed at Penmaenmawr in North Wales and numerous other locations. Rough-outs were made locally near the quarries, and some were polished locally to give a fine finish. This step not only increased the mechanical strength of the axe, but also made penetration of wood easier. Flint was still used from sources such as Grimes Graves but from many other mines across Europe.

Evidence of deforestation has been found in Minoan Crete; for example the environs of the Palace of Knossos were severely deforested in the Bronze Age.

Pre-Industrial History

Easter Island, deforested. According to Jared Diamond: "Among past societies faced with the prospect of ruinous deforestation, Easter Island and Mangareva chiefs succumbed to their immediate concerns, but Tokugawa shoguns, Inca emperors, New Guinea highlanders, and 16th century German landowners adopted a long view and reafforested."

Throughout most of history, humans were hunter gatherers who hunted within forests. In most areas, such as the Amazon, the tropics, Central America, and the Caribbean, only after shortages of wood and other forest products occur are policies implemented to ensure forest resources are used in a sustainable manner.

In ancient Greece, Tjeered van Andel and co-writers summarized three regional studies of historic erosion and alluviation and found that, wherever adequate evidence exists, a major phase of erosion follows, by about 500-1,000 years the introduction of farming in the various regions of Greece, ranging from the later Neolithic to the Early Bronze Age. The thousand years following the mid-first millennium BC saw serious, intermittent pulses of soil erosion in numerous places. The historic silting of ports along the southern coasts of Asia Minor (*e.g.* Clarus, and the examples of Ephesus, Priene and Miletus, where harbors had to be abandoned because of the silt deposited by the Meander) and in coastal Syria during the last centuries BC.

Easter Island has suffered from heavy soil erosion in recent centuries, aggravated by agriculture and deforestation. Jared Diamond gives an extensive look into the collapse of the ancient Easter Islanders in his book Collapse. The disappearance of the island's trees seems to coincide with a decline of its civilization around the 17th and 18th century. He attributed the collapse to deforestation and over-exploitation of all resources.

The famous silting up of the harbor for Bruges, which moved port commerce to Antwerp, also followed a period of increased settlement growth (and apparently of deforestation) in the upper river basins. In early medieval Riez in upper Provence, alluvial silt from two small rivers raised the riverbeds and widened the floodplain, which slowly buried the Roman settlement in alluvium and gradually moved new construction to higher ground; concurrently the headwater valleys above Riez were being opened to pasturage.

A typical progress trap was that cities were often built in a forested area, which would provide wood for some industry (for example, construction, shipbuilding, pottery). When deforestation occurs without proper replanting, however; local wood supplies become difficult to obtain near enough to remain competitive, leading to the city's abandonment, as happened repeatedly in Ancient Asia Minor. Because of fuel needs, mining and metallurgy often led to deforestation and city abandonment.

With most of the population remaining active in (or indirectly dependent on) the agricultural sector, the main pressure in most areas remained land clearing for crop and cattle farming. Enough wild green was usually left standing (and partially used, for example, to collect firewood, timber and fruits, or to graze pigs) for wildlife to remain viable. The elite's (nobility and higher clergy) protection of their own hunting privileges and game often protected significant woodlands.

Major parts in the spread (and thus more durable growth) of the population were played by monastical 'pioneering' (especially by the Benedictine and Commercial orders) and some feudal lords' recruiting farmers to settle (and become tax payers) by offering relatively good legal and fiscal conditions. Even when speculators sought to encourage towns, settlers needed an agricultural belt around or sometimes within defensive walls. When populations were quickly decreased by causes such as the Black Death or devastating warfare (for example, Genghis Khan's Mongol hordes in eastern and central Europe, Thirty

Years' War in Germany), this could lead to settlements being abandoned. The land was reclaimed by nature, but the secondary forests usually lacked the original biodiversity.

Deforestation of Brazil's Atlantic Forest c.1820-1825.

From 1100 to 1500 AD, significant deforestation took place in Western Europe as a result of the expanding human population. The large-scale building of wooden sailing ships by European (coastal) naval owners since the 15th century for exploration, colonisation, slave trade–and other trade on the high seas consumed many forest resources. Piracy also contributed to the over harvesting of forests, as in Spain. This led to a weakening of the domestic economy after Columbus' discovery of America, as the economy became dependent on colonial activities (plundering, mining, cattle, plantations, trade, etc.)

In *Changes in the Land* (1983), William Cronon analyzed and documented 17th-century English colonists' reports of increased seasonal flooding in New England during the period when new settlers initially cleared the forests for agriculture. They believed flooding was linked to widespread forest clearing upstream.

The massive use of charcoal on an industrial scale in Early Modern Europe was a new type of consumption of western forests; even in Stuart England, the relatively primitive production of charcoal has already reached an impressive level. Stuart England was so widely deforested that it depended on the Baltic trade for ship timbers, and looked to the untapped forests of New England to supply the need. Each of Nelson's Royal Navy war ships at Trafalgar (1805) required 6,000 mature oaks for its construction. In France, Colbert planted oak forests to supply the French navy in the future. When the oak plantations matured in the mid-19th century, the masts were no longer required because shipping had changed.

Norman F. Cantor's summary of the effects of late medieval deforestation applies equally well to Early Modern Europe:

Europeans had lived in the midst of vast forests throughout the earlier medieval centuries. After 1250 they became so skilled at deforestation that by 1500 they were running short of wood for heating and cooking. They were faced with a nutritional decline because of the elimination of the generous supply of wild game that had inhabited the now-disappearing forests, which throughout medieval times had provided the staple of their carnivorous high-protein diet. By 1500 Europe was on the edge of a fuel and nutritional disaster [from] which it was saved in the sixteenth century only by the burning of soft coal and the cultivation of potatoes and maize.

INDUSTRIAL ERA

In the 19th century, introduction of steamboats in the United States was the cause of deforestation of banks of major rivers, such as the Mississippi River, with increased and more severe flooding one of the environmental results. The steamboat crews cut wood every day from the riverbanks to fuel the steam engines. Between St. Louis and the confluence with the Ohio River to the south, the Mississippi became more wide and shallow, and changed its channel laterally. Attempts to improve navigation by the use of snag pullers often resulted in crews' clearing large trees 100 to 200 feet (61 m) back from the banks. Several French colonial towns of the Illinois Country, such as Kaskaskia, Cahokia and St. Philippe, Illinois were flooded and abandoned in the late 19th century, with a loss to the cultural record of their archeology.

The wholescale clearance of woodland to create agricultural land can be seen in many parts of the world, such as the Central forest-grasslands transition and other areas of the Great Plains of the United States. Specific parallels are seen in the 20th-century deforestation occurring in many developing nations.

Rates of Deforestation

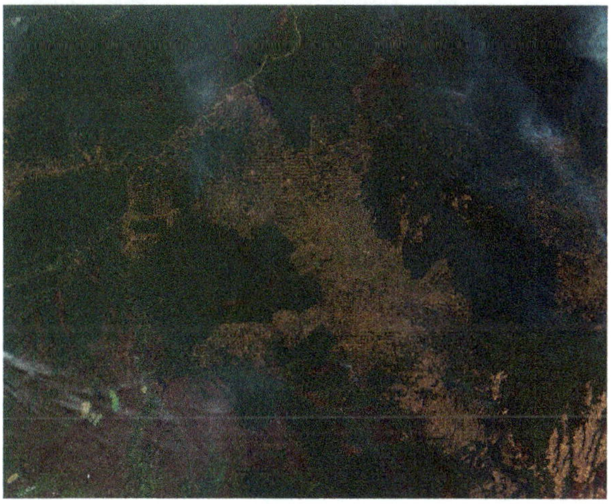

Slash-and-burn farming in the state of Rondônia, western Brazil.

Global deforestation sharply accelerated around 1852. It has been estimated that about half of the Earth's mature tropical forests—between 7.5 million and 8 million km^2 (2.9 million to 3 million sq mi) of the original 15 million to 16 million km^2 (5.8 million to 6.2 million sq mi) that until 1947 covered the planet—have now been destroyed. Some scientists have predicted that unless significant measures (such as seeking out and protecting old growth forests that have not been disturbed) are taken on a worldwide basis, by 2030 there will only be 10% remaining, with another 10% in a degraded condition. 80% will have been lost, and with them hundreds of thousands of irreplaceable species.Some cartographers have attempted to illustrate the sheer scale of deforestation by country using a cartogram. Estimates vary widely as to the extent of tropical deforestation. Scientists estimate that one fifth of the world's tropical rainforest was destroyed between 1960 and 1990. They claim that that rainforests 60 years ago covered 14% of the world's land surface, now only cover 5–7%, and that all tropical forests will be gone by the middle of the 21st century.

A 2002 analysis of satellite imagery suggested that the rate of deforestation in the humid tropics (approximately 5.8 million hectares per year) was roughly 23% lower than the most commonly quoted rates. Conversely, a newer analysis of satellite images reveals that deforestation of the Amazon rainforest is twice as fast as scientists previously estimated. Some have argued that deforestation trends may follow a Kuznets curve, which if true would nonetheless fail to eliminate the risk of irreversible loss of non-economic forest values (for example, the extinction of species).

Satellite image of Haiti's border with the Dominican Republic (right) shows the amount of deforestation on the Haitian side.

A 2005 report by the United Nations Food and Agriculture Organization (FAO) estimates that although the Earth's total forest area continues to decrease at about 13 million hectares per year, the global rate of deforestation has recently been slowing. Still others claim that rainforests are being destroyed at an ever-quickening pace. The London-based Rainforest Foundation notes that "the UN figure is based on a definition of forest as being an area with as little as 10% actual tree cover, which would therefore include areas that are actually savannah-like ecosystems and badly damaged forests."Other critics of the FAO data point out that they do not distinguish between forest types, and that they are based largely on reporting from forestry departments of individual countries, which do not take into account unofficial activities like illegal logging.

Despite these uncertainties, there is agreement that destruction of rainforests remains a significant environmental problem. Up to 90% of West Africa's coastal rainforests have disappeared since 1900. In South Asia, about 88% of the rainforests have been lost. Much of what remains of the world's rainforests is in the Amazon basin, where the Amazon Rainforest covers approximately 4 million square kilometres. The regions with the highest tropical deforestation rate between 2000 and 2005 were Central America—which lost 1.3% of its forests each year—and tropical Asia. In Central America, two-thirds of lowland tropical forests have been turned into pasture since 1950 and 40% of all the rainforests have been lost in the last 40 years. Brazil has lost 90–95% of its Mata Atlântica forest. Paraguay was losing its natural semi humid forests in the country's western regions at a rate of 15.000 hectares at a randomly studied 2-month period in 2010, Paraguay's parliament refused in 2009 to pass a law that would have stopped cutting of natural forests altogether.

Deforestation around Pakke Tiger Reserve, India.

Madagascar has lost 90% of its eastern rainforests. As of 2007, less than 50% of Haiti's forests remained. Mexico, India, the Philippines, Indonesia, Thailand, Burma, Malaysia, Bangladesh, China, Sri Lanka, Laos, Nigeria, the Democratic Republic of the Congo, Liberia, Guinea, Ghana and the Ivory Coast, have lost large areas of their rainforest. Several countries, notably Brazil, have declared their deforestation a national emergency. The World Wildlife Fund's ecoregion project catalogues habitat types throughout the world, including habitat loss such as deforestation, showing for example that even in the rich forests of parts of Canada such as the Mid-Continental Canadian forests of the prairie provinces half of the forest cover has been lost or altered.

Regions

Rates of deforestation vary around the world.

In 2011 Conservation International listed the top 10 most endangered forests, characterized by having all lost 90% or more of their original habitat, and each harboring at least 1500 endemic plant species (species found nowhere else in the world).

Top 10 Most Endangered Forests 2011				
Endangered forest	**Region**	**Remaining habitat**	**Predominate vegetation type**	**Notes**
Indo-Burma	Asia-Pacific	5%	Tropical and subtropical moist broadleaf forests	Rivers, floodplain wetlands, mangrove forests. Burma, Thailand, Laos, Vietnam, Cambodia, India.
New Caledonia	Asia-Pacific	5%	Tropical and subtropical moist broadleaf forests	See note for region covered.
Sundaland	Asia-Pacific	7%	Tropical and subtropical moist broadleaf forests	Western half of the Indo-Malayan archipelago including southern Borneo and Sumatra.
Philippines	Asia-Pacific	7%	Tropical and subtropical moist broadleaf forests	Forests over the entire country including 7,100 islands.
Atlantic Forest	South America	8%	Tropical and subtropical moist broadleaf forests	Forests along Brazil's Atlantic coast, extends to parts of Paraguay, Argentina and Uruguay.

Mountains of Southwest China	Asia-Pacific	8%	Temperate coniferous forest	See note for region covered.
California Floristic Province	North America	10%	Tropical and subtropical dry broadleaf forests	See note for region covered.
Coastal Forests of Eastern Africa	Africa	10%	Tropical and subtropical moist broadleaf forests	Mozambique, Tanzania, Kenya, Somalia.
Madagascar & Indian Ocean Islands	Africa	10%	Tropical and subtropical moist broadleaf forests	Madagascar, Mauritius, Reunion, Seychelles, Comoros.
Eastern Afromontane	Africa	11%	Tropical and subtropical moist broadleaf forests Montane grasslands and shrublands	Forests scattered along the eastern edge of Africa, from Saudi Arabia in the north to Zimbabwe in the south.

CONTROL

Reducing Emissions

Main international organizations including the United Nations and the World Bank, have begun to develop programs aimed at curbing deforestation. The blanket term Reducing Emissions from Deforestation and Forest Degradation (REDD) describes these sorts of programs, which use direct monetary or other incentives to encourage developing countries to limit and/or roll back deforestation. Funding has been an issue, but at the UN Framework Convention on Climate Change (UNFCCC) Conference of the Parties-15 (COP-15) in Copenhagen in December 2009, an accord was reached with a collective commitment by developed countries for new and additional resources, including forestry and investments through international institutions, that will approach USD 30 billion for the period 2010–2012. Significant work is underway on tools for use in monitoring developing country adherence to their agreed REDD targets. These tools, which rely on remote forest monitoring using satellite imagery and other data sources, include the Center for Global Development's FORMA (Forest Monitoring for Action) initiative and the Group on Earth Observations' Forest Carbon Tracking Portal. Methodological guidance for forest monitoring was also emphasized at COP-15. The environmental organization Avoided Deforestation Partners leads the campaign for development of REDD through funding from the U.S. government. In 2014, the Food and Agriculture Organization of the United Nations and partners launched Open Foris - a set of open-source software tools that assist countries in gathering, producing and disseminating information on the state of forest resources. The tools support the inventory lifecycle, from needs assessment, design, planning, field data collection and management, estimation analysis, and dissemination. Remote sensing image processing tools are included, as well as tools for international reporting for Reducing emissions from deforestation and forest degradation (REDD) and MRV and FAO's Global Forest Resource Assessments.

In evaluating implications of overall emissions reductions, countries of greatest concern are those categorized as High Forest Cover with High Rates of Deforestation (HFHD) and Low Forest Cover with High Rates of Deforestation (LFHD). Afghanistan, Benin, Botswana, Burma, Burundi, Cameroon, Chad, Ecuador, El Salvador, Ethiopia, Ghana, Guatemala, Guinea, Haiti, Honduras, Indonesia, Liberia, Malawi, Mali, Mauritania, Mongolia, Namibia, Nepal, Nicaragua, Niger, Nigeria, Pakistan, Paraguay, Philippines, Senegal, Sierra Leone, Sri Lanka, Sudan, Togo, Uganda, United Republic of Tanzania, Zimbabwe are listed as having Low Forest Cover with High Rates of Deforestation (LFHD). Brazil, Cambodia, Democratic Peoples Republic of Korea, Equatorial Guinea, Malaysia, Solomon Islands, Timor-Leste, Venezuela, Zambia are listed as High Forest Cover with High Rates of Deforestation (HFHD).

Payments for Conserving Forests

In Bolivia, deforestation in upper river basins has caused environmental problems, including soil erosion and declining water quality. An innovative project to try and remedy this situation involves landholders in upstream areas being paid by downstream water users to conserve forests. The landholders receive US$20 to conserve the trees, avoid polluting livestock practices, and enhance the biodiversity and forest carbon on their land. They also receive US$30, which purchases a beehive, to compensate for conservation for two hectares of water-sustaining forest for five years. Honey revenue per hectare of forest is US$5 per year, so within five years, the landholder has sold US$50 of honey. The project is being conducted by Fundación Natura Bolivia and Rare Conservation, with support from the Climate & Development Knowledge Network.

Land Rights

Transferring rights over land from public domain to its indigenous inhabitants is argued to be a cost effective strategy to conserve forests. This includes the protection of such rights entitled in existing laws, such as India's Forest Rights Act. The transferring of such rights in China, perhaps the largest land reform in modern times, has been argued to have increased forest cover. In Brazil, forested areas given tenure to indigenous groups have even lower rates of clearing than national parks.

Transferring land rights to indigenous inhabitants is argued to efficiently conserve forests.

Farming

New methods are being developed to farm more intensively, such as high-yield hybrid crops, greenhouse, autonomous building gardens, and hydroponics. These methods are often dependent on chemical inputs to maintain necessary yields. In cyclic agriculture, cattle are grazed on farm land that is resting and rejuvenating. Cyclic agriculture actually increases the fertility of the soil. Intensive farming can also decrease soil nutrients by consuming at an accelerated rate the trace minerals needed for crop growth.The most promising approach, however, is the concept of food forests in permaculture, which consists of agroforestal systems carefully designed to mimic natural forests, with an emphasis on plant and animal species of interest for food, timber and other uses. These systems have low dependence on fossil fuels and agro-chemicals, are highly self-maintaining, highly productive, and with strong positive impact on soil and water quality, and biodiversity.

Monitoring Deforestation

There are multiple methods that are appropriate and reliable for reducing and monitoring deforestation. One method is the "visual interpretation of aerial photos or satellite imagery that is labor-intensive but does not require high-level training in computer image processing or extensive computational resources". Another method includes hot-spot analysis (that is, locations of rapid change) using expert opinion or coarse resolution satellite data to identify locations for detailed digital analysis with high resolution satellite images. Deforestation is typically assessed by quantifying the amount of area deforested, measured at the present time. From an environmental point of view, quantifying the damage and its possible consequences is a more important task, while conservation efforts are more focused on forested land protection and development of land-use alternatives to avoid continued deforestation. Deforestation rate and total area deforested, have been widely used for monitoring deforestation in many regions, including the Brazilian Amazon deforestation monitoring by INPE. A global satellite view is available.

Forest Management

Efforts to stop or slow deforestation have been attempted for many centuries because it has long been known that deforestation can cause environmental damage sufficient in some cases to cause societies to collapse. In Tonga, paramount rulers developed policies designed to prevent conflicts between short-term gains from converting forest to farmland and long-term problems forest loss would cause, while during the 17th and 18th centuries in Tokugawa, Japan, the shoguns developed a highly sophisticated system of long-term planning to stop and even reverse deforestation of the preceding centuries through substituting timber by other products and more efficient use of land that had been farmed for many centuries. In 16th-century Germany, landowners also developed silviculture to deal with the problem of deforestation. However, these policies tend to be limited to environments with *good rainfall*, *no dry season* and *very young* soils (through volcanism or glaciation). This is because on older and less fertile soils trees grow too slowly for silviculture to be economic, whilst in areas with a strong dry season there is always a risk of forest fires destroying a tree crop before it matures.

In the areas where "slash-and-burn" is practiced, switching to "slash-and-char" would prevent the rapid deforestation and subsequent degradation of soils. The biochar thus created, given back to the soil, is not only a durable carbon sequestration method, but it also is an extremely beneficial amendment to the soil. Mixed with biomass it brings the creation of terra preta, one of the richest soils on the planet and the only one known to regenerate itself.

Sustainable Practices

Bamboo is advocated as a more sustainable alternative for cutting down wood for fuel.

Certification, as provided by global certification systems such as Programme for the Endorsement of Forest Certification and Forest Stewardship Council, contributes to tackling deforestation by creating market demand for timber from sustainably managed forests. According to the United Nations Food and Agriculture Organization (FAO), "A major condition for the adoption of sustainable forest management is a demand for products that are produced sustainably and consumer willingness to pay for the higher costs entailed. Certification represents a shift from regulatory approaches to market incentives to promote sustainable forest management. By promoting the positive attributes of forest products from sustainably managed forests, certification focuses on the demand side of environmental conservation." Rainforest Rescue argues that the standards of organizations like FSC are too closely connected to timber industry interests and therefore do not guarantee environmentally and socially responsible forest management. In reality, monitoring systems are inadequate and various cases of fraud have been documented worldwide.

Some nations have taken steps to help increase the amount of trees on Earth. In 1981, China created National Tree Planting Day Forest and forest coverage had now reached 16.55% of China's land mass, as against only 12% two decades ago

Using fuel from bamboo rather than wood results in cleaner burning, and since bamboo matures much faster than wood, deforestation is reduced as supply can be replenished faster.

Reforestation

In many parts of the world, especially in East Asian countries, reforestation and afforestation are increasing the area of forested lands. The amount of woodland has increased in 22 of the world's 50 most forested nations. Asia as a whole gained 1 million hectares of forest between 2000 and 2005. Tropical forest in El Salvador expanded more than 20% between 1992 and 2001. Based on these trends, one study projects that global forest will increase by 10%—an area the size of India—by 2050.

In the People's Republic of China, where large scale destruction of forests has occurred, the government has in the past required that every able-bodied citizen between the ages of 11 and 60 plant three to five trees per year or do the equivalent amount of work in other forest services. The government claims that at least 1 billion trees have been planted in China every year since 1982. This is no longer required today, but March 12 of every year in China is the Planting Holiday. Also, it has introduced the Green Wall of China project, which aims to halt the expansion of the Gobi desert through the planting of trees. However, due to the large percentage of trees dying off after planting (up to 75%), the project is not very successful. There has been a 47-million-hectare increase in forest area in China since the 1970s. The total number of trees amounted to be about 35 billion and 4.55% of China's land mass increased in forest coverage. The forest coverage was 12% two decades ago and now is 16.55%.

An ambitious proposal for China is the Aerially Delivered Re-forestation and Erosion Control System and the proposed Sahara Forest Project coupled with the Seawater Greenhouse.

In Western countries, increasing consumer demand for wood products that have been produced and harvested in a sustainable manner is causing forest landowners and forest industries to become increasingly accountable for their forest management and timber harvesting practices.

The Arbor Day Foundation's Rain Forest Rescue program is a charity that helps to prevent deforestation. The charity uses donated money to buy up and preserve rainforest land before the lumber companies can buy it. The Arbor Day Foundation then protects the land from deforestation. This also locks in the way of life of the primitive tribes living on the forest land. Organizations such as Community Forestry International, Cool Earth, The Nature Conservancy, World Wide Fund for Nature, Conservation International, African Conservation Foundation and Greenpeace also focus on preserving forest habitats. Greenpeace in particular has also mapped out the forests that are still intact and published this information on the internet. World Resources Institute in turn has made a simpler thematic map showing the amount of forests present just before the age of man (8000 years ago) and the current (reduced) levels of forest. These maps mark the amount of afforestation required to repair the damage caused by people.

Forest Plantations

To meet the world's demand for wood, it has been suggested by forestry writers Botkins and Sedjo that high-yielding forest plantations are suitable. It has been calculated that plantations yielding 10 cubic meters per hectare annually could supply all the timber required for international trade on 5% of the world's existing forestland. By contrast, natural forests produce about 1–2 cubic meters per hectare; therefore, 5–10 times more forestland would be required to meet demand. Forester Chad Oliver has suggested a forest mosaic with high-yield forest lands interspersed with conservation land.

In the country of Senegal, on the western coast of Africa, a movement headed by youths has helped to plant over 6 million mangrove trees. The trees will protect local villages from storm damages and will provide a habitat for local wildlife. The project started in 2008, and already the Senegalese government has been asked to establish rules and regulations that would protect the new mangrove forests.

MILITARY CONTEXT

While the preponderance of deforestation is due to demands for agricultural and urban use for the human population, there are some examples of military causes. One example of deliberate deforestation is that which took place in the U.S. zone of occupation in Germany after World War II. Before the onset of the Cold War, defeated Germany was still considered a potential future threat rather than potential future ally. To address this threat, attempts were made to lower German industrial potential, of which forests were deemed an element. Sources in the U.S. government admitted that the purpose of this was that the "ultimate destruction of the war potential of German forests." As a consequence of the practice of clear-felling, deforestation resulted which could "be replaced only by long forestry development over perhaps a century."

American Sherman tanks knocked out by Japanese artillery on Okinawa.

Deforestation can also be one consequence of war. For example, in the 1945 Battle of Okinawa, bombardment and other combat operations reduced the lush tropical landscape into "a vast field of mud, lead, decay and maggots".Deforestation can also be an intentional tactic of military forces. Defoliants (like Agent Orange or others) was used by the British in the Malayan Emergency, and by the United States in the Korean War and Vietnam War.

PUBLIC HEALTH CONTEXT

Deforestation eliminates a great number of species of plants and animals which also often results in an increase in disease. Loss of native species allows new species to come to dominance. Often the destruction of predatory species can result in an increase in rodent populations. These are known to carry plagues. Additionally, erosion can produce pools of stagnant water that are perfect breeding grounds for mosquitos, well known vectors of malaria, yellow fever, nipah virus, and more. Deforestation can also create a path for non-native species to flourish such as certain types of snails, which have been correlated with an increase in schistosomiasis cases.

Deforestation is occurring all over the world and has been coupled with an increase in the occurrence of disease outbreaks. In Malaysia, thousands of acres of forest have been cleared for pig farms. This has resulted in an increase in the zoonosis the Nipah virus. In Kenya, deforestation has led to an increase in malaria cases which is now the leading cause of morbidity and mortality the country.

Another pathway through which deforestation affects disease is the relocation and dispersion of disease-carrying hosts. This disease emergence pathway can be called "range expansion," whereby the host's range (and thereby the range of pathogens) expands to new geographic areas. Through deforestation, hosts and reservoir species are forced into neighboring habitats. Accompanying the reservoir species are pathogens that have the ability to find new hosts in previously unexposed regions. As these pathogens and species come into closer contact with humans, they are infected both directly and indirectly.

A catastrophic example of range expansion is the 1998 outbreak of Nipah Virus in Malaysia. For a number of years, deforestation, drought, and subsequent fires led to a dramatic geographic shift and density of fruit bats, a reservoir for Nipah virus. Deforestation reduced the available fruiting trees in the bats' habitat, and they encroached on surrounding orchards which also happened to be the location of a large number of pigsties. The bats, through proximity spread the Nipah to pigs. While the virus infected the pigs, mortality was much lower than among humans, making the pigs a virulent host leading to the transmission of the virus to humans. This resulted in 265 reported cases of encephalitis, of which 105 resulted in death. This example provides an important lesson for the impact deforestation can have on human health.

Another example of range expansion due to deforestation and other anthropogenic habitat impacts includes the Capybara rodent in Paraguay. This rodent is the host of a number of zoonotic diseases and, while there has not yet been a human-borne outbreak due to the movement of this rodent into new regions, it offers an example of how habitat destruction through deforestation and subsequent movements of species is occurring regularly.

A now well-developed theory is that the spread of HIV it is at least partially due deforestation. Rising populations created a food demand and with deforestation opening up new areas of the forest the hunters harvested a great deal of primate bushmeat, which is believed to be the origin of HIV.

DEFORESTATION AND CLIMATE CHANGE

Deforestation is one of the main causes of climate change. It is the second largest anthropogenic source of carbon dioxide to the atmosphere, after fossil fuel combustion. Deforestation and forest degradation contribute to atmospheric greenhouse gas emissions through combustion of forest biomass and decomposition of remaining plant material and soil carbon. It used to account for more than 20% of carbon dioxide emissions, but it's currently somewhere around the 10% mark. By 2008, deforestation was 12% of total CO_2, or 15% if peatlands are included. These proportions are likely to have fallen since given the continued rise of fossil fuel use.

Averaged over all land and ocean surfaces, temperatures warmed roughly 1.53 °F (0.85 °C) between 1880 and 2012, according to the Intergovernmental Panel on Climate Change. In the Northern Hemisphere, 1983 to 2012 wcrc thc warmcst 30-year period of the last 1400 years.

Effect on climate change

Decrease in Biodiversity

A 2007 study conducted by the National Science Foundation found that biodiversity and genetic diversity are codependent—that diversity among species requires diversity within a species, and vice versa. "If any one type is removed from the system, the cycle can break down, and the community becomes dominated by a single species."

Counteracting climate change

Reforestation

Reforestation is the natural or intentional restocking of existing forests and woodlands that have been depleted, usually through deforestation. It is the reestablishment of forest cover either naturally or artificially. Similar to the other methods of forestation, reforestation can be very effective because a single tree can absorb as much as 48 pounds of carbon dioxide per year and can sequester 1 ton of carbon dioxide by the time it reaches 40 years old.

Afforestation

Afforestation is the establishment of a forest or stand of trees in an area where there was no forest.

China

Although China has set official goals for reforestation, these goals were set for an 80-year time horizon and were not significantly met by 2008. China is trying to correct these problems with projects such as the Green Wall of China, which aims to replant forests and halt the expansion of the Gobi Desert. A law promulgated in 1981 requires that every school student over the age of 11 plant at least one tree per year. But average success rates, especially in state-sponsored plantings, remains relatively low. And even the properly planted trees have had great difficulty surviving the combined impacts of prolonged droughts, pest infestation and fires. Nonetheless, China currently has the highest afforestation rate of any country or region in the world, with 4.77 million hectares (47,000 square kilometers) of afforestation in 2008.

Japan

The primary goal of afforestation projects in Japan is to develop the forest structure of the nation and to maintain the biodiversity found in the Japanese wilderness. The Japanese temperate rainforest is scattered throughout the Japanese archipelago and is home to many endemic species that are not naturally found anywhere else. As development of the country's caused a decline in forest cover, a reduction in biodiversity was seen in those areas.

Agroforestry

Agroforestry or agro-sylviculture is a land use management system in which trees or shrubs are grown around or among crops or pastureland. It combines agricultural and forestry technologies to create more diverse, productive, profitable, healthy, and sustainable land-use systems.

Projects and Foundations

Arbor Day Foundation

Founded in 1972, the centennial of the first Arbor Day observance in the 19th century, the Foundation has grown to become the largest nonprofit membership organization dedicated to planting trees, with over one million members, supporters, and valued partners. They work on projects focused on planting trees around campuses, low-income communities, and communities that have been affected by natural disasters among other places.

Billion Tree Campaign

The **Billion Tree Campaign** *was launched in 2006, by the United Nations Environment* Programme *(UNEP) as a response to the challenges of* global warming, as well as to a wider array of sustainability challenges from water supply to biodiversity *loss. Its initial target was the planting of one billion trees in 2007 (achieved by November). One year later, in 2008, the campaign's objective was raised to 7 billion trees – a target to be met by the climate change conference that was held in* Copenhagen, Denmark *in December 2009. Three months before the conference, the 7 billion planted trees mark had been surpassed. In December 2011, after more than 12 billion trees had been planted, UNEP formally handed management of the program over to the youth-led not-for-profit* Plant-for-the-Planet Foundation, based in Tutzing, Germany. Over 14.2 billion trees have been planted as of 2016. The Billion Tree Campaign calls upon all individuals, countries and companies to participate, and accepts all contributions. Planted trees and pledges can be registered here.

Inspiration

The Billion Tree Campaign was inspired by Nobel Peace Prize *Laureate* Wangari Maathai, founder of the Green Belt Movement. When an executive in the United States told Professor Maathai their corporation was planning to plant a million trees, her response was: "That's great, but what we really need is to plant a billion trees." The campaign was carried out under the patronage of Prince Albert II of Monaco.

"It's the little things citizens do. That's what will make the difference. My little thing is planting trees." – Wangari Maathai

Response

Under UNEP's leadership and through proactive advocacy by the patrons and the partners, the Billion Tree Campaign catalysed tree planting action on all continents. The billionth tree, an African Olive, was planted in Ethiopia in November 2007. The two billionth tree took root as part of the United Nation's World Food Programme agroforestry initiative. The campaign's target was then raised to seven billion trees. In 2009, UNEP mobilized action across the globe through the Twitter for Trees campaign on www.twitter.com/

UNEPandYou. The initiative was driven by a simple yet powerful idea: UNEP pledged to plant one tree to feed into the Billion Tree Campaign, for every follower who joined from 5 May 2009 to World Environment Day on 5 June 2009. The campaign was a success, with 10,300 people following the UNEPandYou page by World Environment Day weekend.

The World Organization of the Scouts Movement *also planted trees under the campaign in line with its mandate to study and protect Nature across several countries.*

The United Nations Peacekeeping missions also joined the campaign and planted trees with its field missions in East-Timor, Côte d'Ivoire, Darfur, Lebanon, Haiti, Congo, and Liberia amongst others missions.

Roll of Honour top 10 countries

- China 2.8 billion
- Pakistan 1 billion
- India 2.1 billion
- Ethiopia 1.6 billion
- Mexico 785 million
- Turkey 716 million
- Nigeria 612 million
- Kenya 455 million
- Peru 246 million
- Myanmar 191 million
- Cuba 137 million

"Around a flowering tree, one finds many insects." – African saying

December 2011, five years since the campaign's launch, the campaign's website www.unep.org/billiontreecampaign proudly registered over 12.5 billion planted trees across 193 countries.

The Amazon Fund (Brazil)

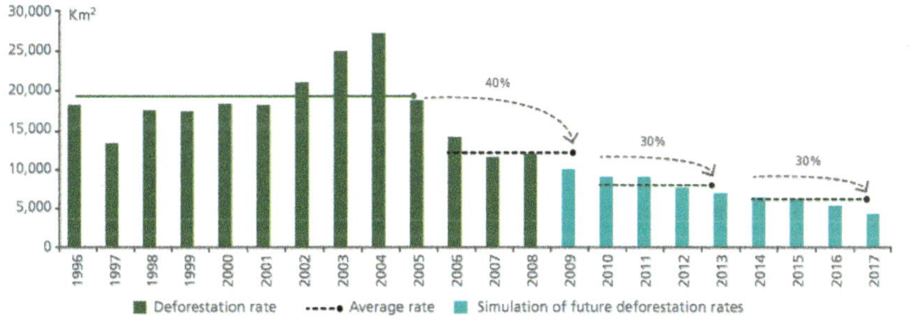

Four-year plan to reduce in deforestation in the Amazon.

Considered the largest reserve of biological diversity in the world, the Amazon Basin is also the largest Brazilian biome, taking up almost half the nation's territory. The Amazon Basin corresponds to two fifths of South America's territory. Its area of approximately seven million square kilometers covers the largest hydrographic network on the planet, through which runs about one fifth of the fresh water on the world's surface. Deforestation in the Amazon rainforest is a major cause to climate change due to the decreasing number of trees available to capture increasing carbon dioxide levels in the atmosphere.

The Amazon Fund is aimed at raising donations for non-reimbursable investments in efforts to prevent, monitor and combat deforestation, as well as to promote the preservation and sustainable use of forests in the Amazon Biome, under the terms of Decree N.° 6,527, dated August 1, 2008. The Amazon Fund supports the following areas: management of public forests and protected areas, environmental control, monitoring and inspection, sustainable forest management, economic activities created with sustainable use of forests, ecological and economic zoning, territorial arrangement and agricultural regulation, preservation and sustainable use of biodiversity, and recovery of deforested areas. Besides those, the Amazon Fund may use up to 20% of its donations to support the development of systems to monitor and control deforestation in other Brazilian biomes and in biomes of other tropical countries.

REFERENCES

1. "Un dizième des terres sauvages ont disparu en deux décennies" (Radio Télévision Suisse) citing Watson, James E.M.; Shanahan, Danielle F.; Di Marco, Moreno; Allan, James; Laurance, William F.; Sanderson, Eric W.; MacKey, Brendan; Venter, Oscar (2016). "Catastrophic Declines in Wilderness Areas Undermine Global Environment Targets". Current Biology. 26 (21): 2929–2934. doi:10.1016/j.cub.2016.08.049. PMID 27618267.
2. SAFnet Dictionary|Definition For [deforestation]. Dictionary of forestry.org (2008-07-29). Retrieved 2011-05-15.
3. Bradford, Alina. (2015-03-04) Deforestation: Facts, Causes & Effects. Livescience.com. Retrieved 2016-11-13.
4. Deforestation | Threats | WWF. Worldwildlife.org. Retrieved 2016-11-13.
5. Oliver, C.D. (1980). "Forest Development in North America following major disturbances". For. Ecol. Management. 3: 153–168. doi:10.1016/0378-1127(80)90013-4.
6. Sahney, S., Benton, M.J. & Falcon-Lang, H.J. (2010). "Rainforest collapse triggered Pennsylvanian tetrapod diversification in Euramerica". Geology. 38 (12): 1079–1082. doi:10.1130/G31182.1.
7. Patel-Weynand, Toral (2002) Biodiversity and sustainable forestry: State of the science review. The National Commission on Science for Sustainable Forestry, Washington DC
8. Kauppi, P. E.; Ausubel, J. H.; Fang, J.; Mather, A. S.; Sedjo, R. A.; Waggoner, P. E. (2006). "Returning forests analyzed with the forest identity". Proceedings of the

National Academy of Sciences. 103 (46): 17574–9. doi:10.1073/pnas.0608343103. PMC 1635979 . PMID 17101996.

9. "Use Energy, Get Rich and Save the Planet", The New York Times, 20 April 2009.
10. Rainforest Facts. Nature.org (2016-11-01). Retrieved 2016-11-13.
11. "Facts About Rainforests". The Nature Conservancy. Retrieved 2015-10-19.
12. UNFCCC (2007). "Investment and financial flows to address climate change" (PDF). unfccc.int. UNFCCC. p. 81.
13. Arild Angelsen; David Kaimowitz (February 1999). "Rethinking the causes of deforestation: Lessons from economic models". The World Bank Research Observer, 14:1. Oxford University Press. pp. 73–98.
14. Laurance, William F. (December 1999). "Reflections on the tropical deforestation crisis" (PDF). Biological Conservation, Volume 91, Issues 2–3. pp. 109–117.
15. Helmut J. Geist And Eric F. Lambin (February 2002). "Proximate Causes and Underlying Driving Forces of Tropical Deforestation" (PDF). BioScience. 52 (2): 143–150. doi:10.1641/0006-3568(2002)052[0143:PCAUDF]2.0.CO;2.
16. Burgonio, T.J. (3 January 2008). "Corruption blamed for deforestation". Philippine Daily Inquirer.
17. "WRM Bulletin Number 74". World Rainforest Movement. September 2003.
18. "Global Deforestation". Global Change Curriculum. University of Michigan Global Change Program. 4 January 2006.
19. Alain Marcoux (August 2000). "Population and deforestation". SD Dimensions. Sustainable Development Department, Food and Agriculture Organization of the United Nations (FAO).
20. Butler, Rhett A. "Impact of Population and Poverty on Rainforests". Mongabay.com / A Place Out of Time: Tropical Rainforests and the Perils They Face. Retrieved 13 May 2009.
21. Stock, Jocelyn; Rochen, Andy. "The Choice: Doomsday or Arbor Day". umich.edu.
22. Karen. "Demographics, Democracy, Development, Disparity and Deforestation: A Crossnational Assessment of the Social Causes of Deforestation". Paper presented at the annual meeting of the American Sociological Association, Atlanta Hilton Hotel, Atlanta, GA, 16 August 2003. Retrieved 13 May 2009.
23. "The Double Edge of Globalization". YaleGlobal Online. Yale University Press. June 2007.
24. Butler, Rhett A. "Human Threats to Rainforests—Economic Restructuring". Mongabay.com / A Place Out of Time: Tropical Rainforests and the Perils They Face. Retrieved 13 May 2009.
25. Susanna B. Hecht; Susan Kandel; Ileana Gomes; Nelson Cuellar; Herman Rosa (2006). "Globalization, Forest Resurgence, and Environmental Politics in El Salvador"(PDF). World Development. 34 (2): 308–323. doi:10.1016/j.worlddev.2005.09.005.

26. Pearce, David W (December 2001). "The Economic Value of Forest Ecosystems" (PDF). Ecosystem Health. 7 (4): 284–296. doi:10.1046/j.1526-0992.2001.01037.x.
27. Erwin H Bulte; Mark Joenje; Hans G P Jansen (2000). "Is there too much or too little natural forest in the Atlantic Zone of Costa Rica?". Canadian Journal of Forest Research. 30 (3): 495–506. doi:10.1139/x99-225.
28. Butler, Rhett A.; Laurance, William F. (August 2008). "New strategies for conserving tropical forests" (PDF). Trends in Ecology & Evolution. 23 (9): 469–472. doi:10.1016/j.tree.2008.05.006.
29. Rudel, T.K. 2005 "Tropical Forests: Regional Paths of Destruction and Regeneration in the Late 20th Century"Columbia University Press ISBN 0-231-13195-X
30. "NASA – Top Story – NASA DATA SHOWS DEFORESTATION AFFECTS CLIMATE".
31. "Massive deforestation threatens food security".
32. Deforestation, ScienceDaily
33. Confirmed: Deforestation Plays Critical Climate Change Role, ScienceDaily, 11 May 2007.
34. Clearing Forests May Transform Local—and Global—Climate; Researchers are finding that massive deforestation may have a profound, and possibly catastrophic, impact on local weather 4 March 2013 Scientific American
35. Deforestation causes global warming, FAO
36. Philip M. Fearnside1 and William F. Laurance, TROPICAL DEFORESTATION AND GREENHOUSE-GAS EMISSIONS, Ecological Applications, Volume 14, Issue 4 (August 2004) pp. 982–986.
37. "Fondation Chirac " Deforestation and desertification".
38. IPCC Fourth Assessment Report, Working Group I Report "The Physical Science Basis", Section 7.3.3.1.5 (p. 527)
39. G.R. van der Werf, D.C.Morton, R.S. DeFries, J.G.J. Olivier, P.S. Kasibhatla, R.B. Jackson, G.J. Collatz and J.T. Randerson (2009). "CO2 emissions from forest loss". Nature Geoscience. 2 (11): 737–738. doi:10.1038/ngeo671.
40. Mumoki, Fiona. "The Effects of Deforestation on our Environment Today." Panorama. TakingITGlobal. 18 July 2006. Web. 24 March 2012.
41. I.C. Prentice. "The Carbon Cycle and Atmospheric Carbon Dioxide" IPCC
42. NASA Data Shows Deforestation Affects Climate In The Amazon. NASA News. 9 June 2004.
43. Findell, Kristen L.; Thomas R. Knutson (2006). "Weak Simulated Extratropical Responses to Complete Tropical Deforestation". Journal of Climate. 19 (12): 2835–2850. doi:10.1175/JCLI3737.1.
44. Sheila Wertz-Kanounnikoff; Laura Ximena Rubio Alvarado. "Why are we seeing "REDD"?". Institute for Sustainable Development and International Relations. Archived from the original on 25 December 2007. Retrieved 2016-11-14.

45. "How can you save the rain forest. 8 October 2006. Frank Field". The Times. London. 8 October 2006. Retrieved 1 April 2010.
46. Broeker, Wallace S. (2006). "Breathing easy: Et tu, O2."Columbia University
47. Moran, Emilio F. (1993). "Deforestation and land use in the Brazilian Amazon". Human Ecology. 21: 1–21. doi:10.1007/BF00890069.
48. R Defries; F Achard; S Brown; M Herold; D Murdiyarso; B Schlamadinger; C Desouzajr (2007). "Earth observations for estimating greenhouse gas emissions from deforestation in developing countries" (PDF). Environmental Science Policy. 10 (4): 385–394. doi:10.1016/j.envsci.2007.01.010.
49. "Underlying Causes of Deforestation". UN Secretary-General's Report.
50. Daniel Rogge. "Deforestation and Landslides in Southwestern Washington". University of Wisconsin-Eau Claire.
51. China's floods: Is deforestation to blame? BBC News. 6 August 1999.
52. The book "Environment", 5th Edition, Raven and Berg, John Wiley & Sons, 2006, page 406.
53. Hongchang, Wang (1998-01-01). "Deforestation and Desiccation in China A Preliminary Study". In Schwartz, Jonathan Matthew. The Economic Costs of China's Environmental Degradation: Project on Environmental Scarcities, State Capacity, and Civil Violence, a Joint Project of the University of Toronto and the American Academy of Arts and Sciences. Committee on Internat. Security Studies, American Acad. of Arts and Sciences.
54. "Soil, Water and Plant Characteristics Important to Irrigation". North Dakota State University.
55. Sten Nilsson, Do We Have Enough Forests?, American Institute of Biological Sciences, March 2001.
56. Rainforest Biodiversity Shows Differing Patterns, ScienceDaily, 14 August 2007.
57. "Medicine from the rainforest". Research for Biodiversity Editorial Office. Archived from the original on 6 December 2008.
58. Single-largest biodiversity survey says primary rainforest is irreplaceable, Bio-Medicine, 14 November 2007.
59. Tropical rainforests – The tropical rainforest, BBC
60. Tropical Rain Forest. thinkquest.org
61. U.N. calls on Asian nations to end deforestation, Reuters, 20 June 2008.
62. "Rainforest Facts".
63. Tropical rainforests – Rainforest water and nutrient cycles, BBC
64. Butler, Rhett A. (2 July 2007) Primary rainforest richer in species than plantations, secondary forests, mongabay.com,
65. Flowers, April. "Deforestation In The Amazon Affects Microbial Life As Well As Ecosystems". Science News. Redorbit.com. Retrieved 12 March 2013.
66. Rainforest Facts. Rain-tree.com (2010-03-20). Retrieved 2010-08-29.

67. Leakey, Richard and Roger Lewin, 1996, The Sixth Extinction : Patterns of Life and the Future of Humankind, Anchor, ISBN 0-385-46809-1.
68. The great rainforest tragedy, The Independent, 28 June 2003.
69. Biodiversity wipeout facing South East Asia, New Scientist, 23 July 2003.
70. Pimm, S. L.; Russell, G. J.; Gittleman, J. L.; Brooks, T. M. (1995). "The Future of Biodiversity". Science. 269 (5222): 347–350. doi:10.1126/science.269.5222.347. PMID 17841251.
71. Pimm Stuart L, Russell Gareth J, Gittleman John L, Brooks Thomas M (1995). "The future of biodiversity". Science. 269 (5222): 347–341. doi:10.1126/science.269.5222.347. PMID 17841251.
72. Timothy Charles Whitmore; Jeffrey Sayer; International Union for Conservation of Nature and Natural Resources. General Assembly; IUCN Forest Conservation Programme (15 February 1992). Tropical deforestation and species extinction. Springer. ISBN 978-0-412-45520-9. Retrieved 4 December 2011.
73. Sohn, Emily (12 July 2012). "More extinctions expected in Amazon". Discovery.
74. Nature loss 'to hurt global poor', BBC News, 29 May 2008.
75. Forest Products. (PDF). Retrieved 2011-12-04.
76. "Destruction of Renewable Resources". rainforests.mongabay.com.
77. Deforestation Across the World's Tropical Forests Emits Large Amounts of Greenhouse Gases with Little Economic Benefits, According to a New Study at CGIAR.org, 4 December 2007.
78. "New ASB Report finds deforestation offers very little money compared to potential financial benefits". asb.cgiar.org.
79. Chomitz, Kenneth; Gray, David A. (1999). "Roads, lands, markets, and deforestation : a spatial model of land use in Belize". Policy Research Working Papers. doi:10.1596/1813-9450-1444.
80. Ferraz, Silvio Frosini de Barros; Vettorazzi, Carlos Alberto; Theobald, David M. (2009). "Using indicators of deforestation and land-use dynamics to support conservation strategies: A case study of central Rondônia, Brazil". Forest Ecology and Management. 257 (7): 1586–1595. doi:10.1016/j.foreco.2009.01.013.
81. "Stolen Goods: The EU's complicity in illegal tropical deforestation" (PDF). Forests and the European Union Resource Network. 17 March 2015. Retrieved 31 March 2015.
82. "Reducing Emissions from Deforestation and Forest Degradation (REDD): An Options Assessment Report"(PDF). The Government of Norway. 28 March 2010. p. 16.
83. Meyfroidt, P., Lambin, E.F. 2011. Global Forest Transition: Prospects for an End to Deforestation. Annual Review of Environment and Resources 36: 343-371.
84. Taylor, Leslie (2004). The Healing Power of Rainforest Herbs: A Guide to Understanding and Using Herbal Medicinals. Square One. ISBN 9780757001444.
85. Flannery, T (1994). The future eaters. Melbourne: Reed Books. ISBN 0-7301-

0422-2.

86. Brown, Tony (1997). "Clearances and Clearings: Deforestation in Mesolithic/ Neolithic Britain". Oxford Journal of Archaeology. 16 (2): 133–146. doi:10.1111/1468-0092.00030.
87. "hand tool: Neolithic tools". Encyclopædia Britannica Online.
88. "Neolithic Age from 4,000 BC to 2,200 BC or New Stone Age".
89. Hogan, C. Michael (22 December 2007). "Knossos fieldnotes", The Modern Antiquarian
90. Jared Diamond, Collapse: How Societies Choose to Fail or Survive, chapter "The world as a polder: what does it all mean to us today?", section "Reasons for hope", Penguin Books, 2005 and 2011, pages 522–523 (ISBN 978-0-241-95868-1).
91. Encyclopédia Britannica Online School Edition. School.eb.com. Retrieved 2010-08-29.
92. Tjeerd H. van Andel; Eberhard Zangger; Anne Demitrack (2013). "Land Use and Soil Erosion in Prehistoric and Historical Greece" (PDF). Journal of Field Archaeology. 17 (4): 379–396. doi:10.1179/009346990791548628.
93. "The Mystery of Easter Island", Smithsonian Magazine, 1 April 2007.
94. "Historical Consequences of Deforestation: Easter Island".
95. "Jared Diamond, Easter Island's End".
96. Chew, Sing C. (2001). World Ecological Degradation. Oxford, England: AltaMira Press. pp. 69–70.
97. Norman F. Cantor (9 June 1994). The civilization of the Middle Ages: a completely revised and expanded edition of Medieval history, the life and death of a civilization. HarperCollins. p. 564. ISBN 978-0-06-092553-6. Retrieved 4 December 2011.
98. F. Terry Norris, "Where Did the Villages Go? Steamboats, Deforestation, and Archaeological Loss in the Mississippi Valley", in Common Fields: an environmental history of St. Louis, Andrew Hurley, ed., St. Louis, MO: Missouri Historical Society Press, 1997, ISBN 978-1-883982-15-7 pp. 73–89.
99. Duke Press policy studies / Global deforestation and the nineteenth-century world economy / edited by Richard P. Tucker and J. F. Richards
100. E. O. Wilson, 2002, The Future of Life, Vintage ISBN 0-679-76811-4.
101. Map reveals extent of deforestation in tropical countries, guardian.co.uk, 1 July 2008.
102. Maycock, Paul F. Deforestation. WorldBookOnline.
103. Ron Nielsen, The Little Green Handbook: Seven Trends Shaping the Future of Our Planet, Picador, New York (2006) ISBN 978-0-312-42581-4.
104. Victor Vescovo. (2006). The Atlas of World Statistics. The Caladan Press. Retrieved 2012-08-03.

105. Teja Tscharntke, Christoph Leuschner, Edzo Veldkamp, Heiko Faust, Edi Guhardja, eds. (2010). Tropical Rainforests and Agroforests Under Global Change. Springer. pp. 270–271. ISBN 978-3-642-00492-6.
106. Intergovernmental Panel on Climate Change (2000). Land Use, Land Use Change and Forestry. Cambridge University Press.
107. John F. Mongillo; Linda Zierdt-Warshaw (2000). Linda Zierdt-Warshaw, ed. Encyclopedia of environmental science. University of Rochester Press. p. 104. ISBN 978-1-57356-147-1.
108. Achard Frederic, Eva Hugh D, Hans- , Stibig Jurgen, Mayaux Philippe (2002). "Determination of deforestation rates of the world's humid tropical forests". Science. 297 (5583): 999–1003. doi:10.1126/science.1070656. PMID 12169731.
109. Jha, Alok (21 October 2005). "Amazon rainforest vanishing at twice rate of previous estimates". The Guardian.
110. Satellite images reveal Amazon forest shrinking faster, csmonitor.com, 21 October 2005.
111. Culas, Richard J. (2007). "Deforestation and the environmental Kuznets curve: An institutional perspective"(PDF). Ecological Economics. 61 (2–3): 429–437. doi:10.1016/j.ecolecon.2006.03.014.
112. Environmental Economics: A deforestation Kuznets curve?, 22 November 2006.
113. "Is there an environmental Kuznets curve for deforestation?".
114. "Pan-tropical Survey of Forest Cover Changes 1980–2000". Forest Resources Assessment. Rome, Italy: Food and Agriculture Organization of the United Nations (FAO).
115. Committee On Forestry. FAO (2001-03-16). Retrieved 2010-08-29.
116. Worldwatch: Wood Production and Deforestation Increase & Recent Content, Worldwatch Institute
117. Rhett A. Butler (16 November 2005). "World deforestation rates and forest cover statistics, 2000–2005". mongabay.com.
118. The fear is that highly diverse habitats, such as tropical rainforest, are vanishing at a faster rate that is partly masked by the slower deforestation of less biodiverse, dry, open forests. Because of this omission, the most harmful impacts of deforestation (such as habitat loss) could be increasing despite a possible decline in the global rate of deforestation.
119. "Remote sensing versus self-reporting".
120. The World Bank estimates that 80% of logging operations are illegal in Bolivia and 42% in Colombia, while in Peru, illegal logging accounts for 80% of all logging activities. (World Bank (2004). Forest Law Enforcement.) (The Peruvian Environmental Law Society (2003). Case Study on the Development and Implementation of Guidelines for the Control of Illegal Logging with a View to Sustainable Forest Management in Peru.)
121. "Forest Holocaust". National Geographic.

122. "Rainforests & Agriculture".
123. The Amazon Rainforest, BBC, 14 February 2003.
124. John Revington. "The Causes of Tropical Deforestation". New Renaissance Magazine.
125. "What is Deforestation?". kids.mongabay.com.
126. "Paraguay es principal deforestador del Chaco". ABC Color newspaper, Paraguay. Retrieved 13 August 2011.
127. "Paraguay farmland". Retrieved 13 August 2011.
128. IUCN – Three new sites inscribed on World Heritage List, 27 June 2007.
129. "Madagascar's rainforest map". New Scientist.
130. "Haiti Is Covered with Trees". EnviroSociety. Tarter, Andrew. Retrieved 14 November 2016.
131. Chart – Tropical Deforestation by Country & Region. Mongabay.com. Retrieved 2011-12-04.
132. Rainforest Destruction. rainforestweb.org
133. Amazon deforestation rises sharply in 2007, USATODAY.com, 24 January 2008.
134. Vidal, John (31 May 2005). "Rainforest loss shocks Brazil". The Guardian. London. Retrieved 1 April 2010.
135. The World's 10 Most Threatened Forest Hotspots, Conservation International, 2 February 2011.
136. Indo-Burma, Conservation International.
137. New Caledonia, Conservation International.
138. Sundaland, Conservation International.
139. Philippines, Conservation International.
140. Atlantic Forest Archived 12 December 2011 at the Wayback Machine., Conservation International.
141. Mountains of Southwest China, Conservation International.
142. California Floristic Province Archived 14 April 2011 at the Wayback Machine., Conservation International.
143. Coastal Forests of Eastern Africa, Conservation International.
144. Madagascar & Indian Ocean Islands, Conservation International.
145. Eastern Afromontane, Conservation International.
146. "Copenhagen Accord of 18 December 2009" (PDF). UNFCC. 2009. Retrieved 2009-12-28.
147. Forest Monitoring for Action (FORMA) : Center for Global Development : Initiatives: Active. Cgdev.org (2009-11-23). Retrieved 2010-08-29.
148. Browser – GEO FCT Portal. Portal.geo-fct.org. Retrieved 2010-08-29.
149. "Methodological Guidance" (PDF). UNFCC. 2009. Retrieved 2009-12-28.

150. Agriculture Secretary Vilsack: $1 billion for REDD+ " Climate Progress. Climateprogress.org (2009-12-16). Retrieved 2010-08-29.
151. "FAO sets standards to improve national forest monitoring systems".
152. Angelsen, Arild; et al. (2009). "Reducing Emissions from Deforestation and Forest Degradation (REDD): An Options Assessment Report" (PDF). Meridian Institute for the Government of Norway. pp. 75–77. Retrieved 2011-11-24.
153. Payments for watershed services: A driver of climate compatible development, Climate & Development Knowledge Network, 30 December 2013.
154. "India should follow China to find a way out of the woods on saving forest people". The Guardian. 22 July 2016. Retrieved 7 August 2016.
155. "China's forest tenure reforms". rightsandresources.org. Retrieved 7 August 2016.
156. "Global Forest Change - Google Crisis Map". Google Crisis Map. Retrieved 12 October 2016.
157. "Warning to forest destroyers: this scientist will catch you". Nature News & Comment. 4 October 2016. Retrieved 12 October 2016.
158. Diamond, Jared Collapse: How Societies Choose To Fail or Succeed; Viking Press 2004, pp. 301–302 ISBN 0-14-311700-9.
159. Diamond, Jared Collapse: How Societies Choose To Fail or Succeed; Viking Press 2004, pp. 320–331 ISBN 0-14-311700-9.
160. Rosenberg, Tina (2012-03-13). "In Africa's vanishing forests, the benefits of bamboo". New York Times. Retrieved 2012-07-26.
161. "State of the World's Forests 2009". United Nations Food and Agriculture Organization.
162. Facts about Tropical Timber. Rainforest Rescue. Retrieved 2016-11-13.
163. John Gittings. "Battling China's Deforestation." The Guardian. 20 March 2001.
164. Foley Jonathan A; DeFries Ruth; Asner Gregory P; Barford Carol; et al. (2005). "Global Consequences of Land Use". Science. 309 (5734): 570–574. doi:10.1126/science.1111772. PMID 16040698.
165. James Owen, "World's Forests Rebounding, Study Suggests" National Geographic News, 13 November 2006.
166. John Gittings, "Battling China's deforestation", World News, 20 March 2001.
167. The world's last intact forest landscapes. intactforests.org
168. "World Intact Forests campaign by Greenpeace". intactforests.org.
169. The World's Forests from a Restoration Perspective, WRI
170. "Alternative thematic map by Howstuffworks; in pdf"(PDF).
171. Daniel B. Botkin (2001). No man's garden: Thoreau and a new vision for civilization and nature. Island Press. pp. 246–247. ISBN 978-1-55963-465-6. Retrieved 4 December 2011.
172. Stenstrup, Allen (2010). Forests. Greensboro, North Carolina: Morgan Reynolds Publishing. p. 89. ISBN 978-1-59935-116-2.

173. Balabkins, Nicholas (1964) "Germany Under Direct Controls; Economic Aspects Of Industrial Disarmament 1945–1948, Rutgers University Press. p. 119. The two quotes used by Balabkins are referenced to, respectively: U.S. office of Military Government, A Year of Potsdam: The German Economy Since the Surrender (1946), p. 70; and U.S. Office of Military Government, The German Forest Resources Survey (1948), p. II. For similar observations see G.W. Harmssen, Reparationen, Sozialproduct, Lebensstandard (Bremen: F. Trujen Verlag, 1948), I, 48.

174. Higa, Takejiro. Battle of Okinawa, The Hawaii Nisei Project

175. Pesticide Dilemma in the Third World: A Case Study of Malaysia. Phoenix Press. 1984. p. 23.

176. "Encyclopedia of World Environmental History". Routledge, 2004. ISBN 0-415-93733-7.

177. Marchak, M. Patricia (18 September 1995). Logging the globe. McGill-Queen's Press – MQUP. pp. 157–. ISBN 978-0-7735-1346-4. Retrieved 4 December 2011.

178. Biodiversity and Infectious Diseases | The Center for Health and the Global Environment. Chgeharvard.org (2011-11-11). Retrieved 2016-11-13.

179. Forests and human health. Fao.org. Retrieved 2016-11-13.

180. Moslemi, Jennifer M.; Snider, Sunny B.; MacNeill, Keeley; Gilliam, James F.; Flecker, Alexander S. (2012). "Impacts of an Invasive Snail". PLoS ONE. 7 (6): e38806. doi:10.1371/journal.pone.0038806. PMC 3382606 . PMID 22761706.

181. Deforestation and emerging diseases | Bulletin of the Atomic Scientists. Thebulletin.org (2011-02-15). Retrieved 2016-11-13.

182. African Politics Portal | Tag Archive | Environmental impact of deforestation in Kenya. African-politics.com (2009-05-28). Retrieved 2016-11-13.

183. 2014 Kenya Economic Survey Marks Malaria As Country's Leading Cause Of Death | The Henry J. Kaiser Family Foundation. Kff.org (2014-05-01). Retrieved 2016-11-13.

184. Scheidt, Spencer N.; Hurlbert, Allen H. (2014). "Range Expansion and Population Dynamics of an Invasive Species: The Eurasian Collared-Dove (Streptopelia decaocto)". PLoS ONE. 9 (10): e111510. doi:10.1371/journal.pone.0111510. PMC 4213033 . PMID 25354270.

185. Lam, Sai Kit; Chua, Kaw Bing (2002). "Nipah Virus Encephalitis Outbreak in Malaysia". Clinical Infectious Diseases. 34: S48–51. doi:10.1086/338818. PMID 11938496.

186. "Nipah Virus (NiV)". cdc.gov.

187. Deforestation sparks giant rodent invasions. News.mongabay.com (2010-12-15). Retrieved 2016-11-13.

CHAPTER 5

Slash-and-Burn Technique

Chapter Contents

- **Introduction**
- **History**
- **Historical References**
- **Ecological Implications**

Slash and burn is a forestry technique that involves cutting down trees and shrubs on a large swath of land, allowing them to dry, and then setting a fire. It is a rapid and effective method for clearing a chunk of land, typically with the intention of converting it to agricultural use. This land clearing method is also cited as a leading cause of deforestation worldwide, since it is extremely destructive when it is not practiced sustainably.

Although many people think of Amazonian forests when they visualize slash and burn, the practice is actually widespread around the world and quite old. Humans have been using fire to clear land for thousands of years, although early methods were practiced on a much smaller scale. Many European forests bear evidence of former slash and burn practices, and it is not uncommon in Africa and Australia as well.

INTRODUCTION

Slash-and-burn agriculture, or fire–fallow cultivation, is a farming method that involves the cutting and burning of plants in a forest or woodland to create a field called a *swidden*. (Preparing fields by deforestation is called *assarting*.) In subsistence agriculture, slash-and-burn typically uses little technology. It is often applied in shifting cultivation agriculture (such as in the Amazon rainforest) and in transhumance livestock herding.

Slash-and-burn is used by 200–500 million people worldwide. In 2004 it was estimated that in Brazil alone, 500,000 small farmers each cleared an average of one hectare (2.47105 acres) of forest per year. The technique is not scalable or sustainable for large human populations. Methods such as Inga alley farming have been proposed as alternatives to this environmental degradation.

HISTORY

Historically, slash-and-burn cultivation has been practiced throughout much of the world, in grasslands as well as woodlands.

During the Neolithic Revolution, which included agricultural advancements, groups of hunter-gatherers domesticated various plants and animals, permitting them to settle down and practice agriculture, which provides more nutrition per hectare than hunting and gathering. This happened in the river valleys of Egypt and Mesopotamia. Due to this decrease in food from hunting, as human populations increased, agriculture became more important. Some groups could easily plant their crops in open fields along river valleys, but others had forests blocking their farming land.

In this context, humans used slash-and-burn agriculture to clear more land to make it suitable for plants and animals. Thus, since Neolithic times, slash-and-burn techniques have been widely used for converting forests into crop fields and pasture. Fire was used before the Neolithic as well, and by hunter-gatherers up to present times. Clearings created by fire were made for many reasons, such as to draw game animals and to promote certain kinds of edible plants such as berries.

Slash-and-burn fields are typically used and owned by a family until the soil is exhausted. At this point the ownership rights are abandoned, the family clears a new field, and trees and shrubs are permitted to grow on the former field. After a few decades, another family or clan may then use the land and claim usufructuary rights. In such a system there is typically no market in farmland, so land is not bought or sold in the open market and land rights are traditional. In slash-and-burn agriculture, forests are typically cut months before a dry season. The "slash" is permitted to dry, and then burned in the following dry season. The resulting ash fertilizes the soil and the burned field is then planted at the beginning of the next rainy season with crops such as upland rice, maize, cassava, or other staples. Most of this work is typically done by hand, using such basic tools as machetes, axes, hoes, and makeshift shovels.

Large families or clans wandering in the lush woodlands long continued to be the most common form of life through human prehistory. Axes to fell trees and sickles for harvesting grain were the only tools people might bring with them. All other tools were made from materials they found at the site, such as fire stakes of birch, long rods (vanko), and harrows made of spruce tops. The extended family conquered the lush virgin forest, burned and cultivated their carefully selected swidden plots, sowed one or more crops, and then proceeded on to forests that had been noted in their wanderings. In the temperate zone the forest regenerated in the course of a lifetime. So swidden was repeated several times in the same area over the years. But in the tropics the forest floor gradually depleted. It was not only in the moors, as in Northern Europe, but also in the steppe, savannah, prairie, pampas and barren desert in tropical areas where shifting cultivation is the oldest type of farming (Clark 1952 91–107).

HISTORICAL REFERENCES

Painting by Eero Järnefelt of forest-burning.

Southern European Mediterranean climates have favored evergreen and deciduous forests. With slash-and-burn agriculture, this type of forest was less able to regenerate than those north of the Alps. Although in northern Europe one crop was usually harvested before grass was allowed to grow, in southern Europe it was more common to exhaust the soil by farming it for several years.

Classical authors mentioned large forests, with Homer writing about "wooded Samothrace," Zakynthos, Sicily, and other woodlands. These authors indicated that the Mediterranean area once had more forest; much had already been lost, and the remainder was primarily in the mountains.

Although parts of Europe aside from the north remained wooded, by the Roman Iron and early Viking Ages, forests were drastically reduced and settlements regularly moved. The reasons for this pattern of mobility, the transition to stable settlements from the late Viking period on, or the transition from shifting cultivation to stationary farming are unknown. From this period, plows are found in graves. Early agricultural peoples preferred good forests on hillsides with good drainage, and traces of cattle enclosures are evident there.

Greek explorer and merchant Pytheas of Marseilles made a voyage to Northern Europe around 330 BC, with part of his itinerary recorded by Polybios, Pliny, and Strabo. Pytheas visited Thule, a six-day voyage north of Britain: "The barbarians showed us the place where the sun does not go to sleep. It happened because there the night was very short—in some places two, in others three hours—so that the sun shortly after its fall soon went up again." He describes a fertile land, "rich in fruits that were ripe only until late in the year, and the people there used to prepare a drink of honey. And they threshed the grain in large houses because of the cloudy weather and frequent rain. In the spring they drove the cattle up into the mountain pastures and stayed there all summer."

In Italy, shifting cultivation was a thing of the past by the birth of Christ. Tacitus describes it as a strange cultivation method, practiced by the Germans. In 98 AD, he wrote about the Germans that their fields were proportional to the participating cultivators but their crops were shared according to status. Distribution was simple, because of wide availability; they changed fields annually, with much to spare because they were producing grain rather than other crops. According to the original text, "Agri pro numero cultorum ab universis in vices occupantur, quos mox inter se secundum dignationem partiuntur, facilitatem partiendi camporum spatia praestant. Arva per annos mutant, et superest ager; nec enim cum ubertate et amplitudine soli labore contendunt, ut pomaria conserant et prata separent et hortos rigent; sola terrae seges imperatur." This is the practice of shifting cultivation.

During the Migration Period in Europe, after the Roman Empire and before the Viking Age, the peoples of Central Europe moved to new forests after exhausting old parcels. Forests were quickly exhausted; the practice had ended in the Mediterranean, where forests were less resilient than the sturdier coniferous forests of Central Europe. Deforestation had been partially caused by burning to create pasture. Reduced timber delivery led to higher prices and more stone construction in the Roman Empire (Stewart 1956, p. 123). Although forests gradually decreased in northern Europe, they have survived in the Nordic countries.

Tribes in pre-Roman Italy (including the Etruscans, Umbrians, Ligurians, Sabines, Latins, Campanians, Apulians, Saliscans, and Sabellians) apparently lived in temporary locations. They cultivated small patches of land, kept sheep and cattle, traded with foreign merchants, and occasionally fought. These Italic groups developed identities as settlers and warriors around 900 BC. They built forts in the mountains which are studied today, as are the ruins of a large Samnite temple and theater at Pietrabbondante.

Many Italic peoples saw benefits in allying with Rome. When the Romans built the *Via Amerina* in 241 BC, the Falisci settled in cities on the plains and aided the Romans in road construction; the Roman Senate gradually acquired representatives from Faliscan and Etruscan families, and the Italic tribes became settled farmers.

Classical writers described peoples who practiced shifting cultivation, which characterized the Migration Period in Europe. The exploitation of forests demanded displacement as areas were deforested. Julius Caesar wrote about the Suebi in *Commentarii de Bello Gallico* 4.1, "They have no private and secluded fields ("*privati ac separati agri apud eos nihil est*") ... They cannot stay more than one year in a place for cultivation's sake" ("*neque longius anno remanere uno in loco colendi causa licet*"). The Suebi lived between the Rhine and the Elbe. About the Germani, Caesar wrote: "No one has a particular field or area for himself, for the magistrates and chiefs give year by year to the people and the clans, who have gathered together, as much land and in such places as seem good to them and then make them move on after a year" ("*Neque quisquam agri modum certum aut fines habet proprios, sed magistratus ac principes in annos singulos gentibus cognationibusque hominum, qui tum una coierunt, a quantum et quo loco visum est agri attribuunt atque anno post alio transire cogunt*" [Book 6.22]).

Strabo (63 BC—c. 20 AD) also writes about the Suebi in his *Geography* (VII, 1, 3): "Common to all the people in this area is that they can easily change residence because of their sordid way of life; they do not cultivate fields or collect property, but live in temporary huts. They get their nourishment from their livestock for the most part, and like nomads, pack all their goods in wagons and go on to wherever they want". Horace writes in 17 BC (*Carmen Saeculare*, 3, 24, 9ff.) about the people of Macedonia: "The proud Getae also live happily, growing free food and cereal for themselves on land they do not want to cultivate for more than a year" ("*Vivunt et rigidi Getae, / immetata quibus iugera liberas / fruges et Cererem ferunt, / nec cultura placet longior annua*").

Locations of Norwegian tribes described by Jordanes in his *Getica*

Jordanes, of Gothic descent, became a monk in Italy. In his mid-sixth-century AD *Getica* (*De origine actibusque Getarum*; *The Origin and Deeds of the Goths*), he described the large island of Scandza, on which the Goths originated. According to Jordanes, of the tribes living there, some are Adogit from within 40 days of the midnight sun. After the Adogit were the Screrefennae and Suehans, who also lived in the north. The Screrefennae did not raise crops, instead hunting and collecting bird eggs. The Suehans, a semi-nomadic tribe with good horses (comparable to the Thuringii), hunted furs to sell; grain could not be grown so far north. In about 550 AD, Procopius also described a primitive hunting people he called "Skrithifinoi": "Both men and women engaged incessantly just in hunting the rich forests and mountains, which gave them an endless supply of game and wild animals."

Slash-and-burn in Småland, Sweden (1904)

Adam of Bremen described Sweden from information he received from the Danish king Sven Estridson (also called Sweyn II of Denmark) in 1068: "It is very fruitful, the earth holds many crops and honey, it has greater livestock than all other countries, there are many useful rivers and forests; with regard to women they do not know moderation; they have for their households two, three, or more wives simultaneously; the rich and the rulers are innumerable ... [with] livestock grazing, as with the Arabs, far out in the wilderness". The use of fire in northeastern Sweden changed as agriculture evolved. Although the Sami people did not burn land (since burning killed the lichen required by their reindeer), later farmers frequently used slash-and-burn techniques. The 19th-century Swedish timber industry moved north, clearing the land of trees but leaving waste behind as a fire risk; during the 1870s, fires were frequent. There was a fire in Norrland in 1851, followed by fires in 1868 and 1878; two towns were lost in 1888.

Forest Finns

Huuhta cultivation spread: within the circle in 1500 AD, within the line in 1600, and to the dashed line in 1700.

One culture which flourished in pre-agricultural Europe survives: the Forest Finns in Scandinavia. Martin Tvengsberg, a descendant the Forest Finns, studied them in his capacity as curator of the Hedmark Museum in Norway. The Savo-Karelians had a sophisticated system for cultivating spruce forests. A runic poem about Finland's spruce forests reads, *«Gåivu on mehdien valgoinen valhe"* ("The birch is the forest's white lie"). The best spruce forests reportedly contain birch trees, which grow only after a forest has burned once or twice.

Modern Western world

Slash-and-burn may be defined as the large-scale deforestation of forests for agricultural use. Ashes from the trees help farmers by providing nutrients for the soil.

In industrialized regions, including Europe and North America, the practice was abandoned with the introduction of market agriculture and land ownership. Slash-and-burn agriculture was initially practiced by European pioneers in North America such as Daniel Boone and his family, who cleared land in the Appalachian Mountains during the late 18th and early 19th centuries. However, land cleared by slash-and-burn farmers was eventually taken over by systems of land tenure focusing on long-term improvement and discouraging practices associated with slash-and-burn agriculture.

Northern European heritage

Telkkämäki Heritage Farm and Nature Reserve in Kaavi, Finland

Some areas of the reserve are burned annually in Telkkämäki, Finland.

Telkkämäki Nature Reserve in Kaavi, Finland, is an open-air museum which still practices slash-and-burn agriculture. Farm visitors can see how people farmed when slash-and-burn agriculture became the norm in the Northern Savonian region of eastern Finland beginning in the 15th century. Areas of the reserve are burnt each year.

South Asia

Tribal groups in the northeastern Indian states of Arunachal Pradesh, Meghalaya, Mizoram and Nagaland and the Bangladeshi districts of Rangamati, Khagrachari, Bandarban and Sylhet refer to slash-and-burn agriculture as *jhum* or *jhoom* cultivation. The system involves clearing land, by fire or clear-felling, for economically-important crops such as upland rice, vegetables or fruits. After a few cycles, the land's fertility declines and a new area is chosen. *Jhum* cultivation is most often practiced on the slopes of thickly-forested hills. Cultivators cut the treetops to allow sunlight to reach the land, burning the trees and grasses for fresh soil. Although it is believed that this helps fertilize the land, it can leave it vulnerable to erosion. Holes are made for the seeds of crops such as sticky rice, maize, eggplant and cucumber are planted. After considering *jhum*'s effects, the government of Mizoram has introduced a policy to end the method in the state. Slash-and-burn is typically a type of subsistence agriculture, not focused on a need to sell crops globally; planting decisions are governed by the needs of the family (or clan) for the coming year.

ECOLOGICAL IMPLICATIONS

Modern Slash-and-Burn Practice

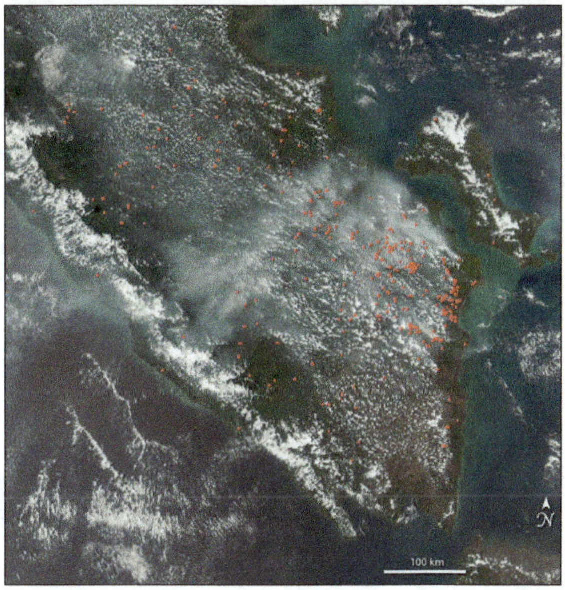

Sumatra, Indonesia.

Chiang Mai, Thailand.

Santa Cruz, Bolivia.

Morondava, Madagascar.

Although a solution for overpopulated tropical countries where subsistence agriculture may be the traditional method of sustaining many families, the consequences of slash-and-burn techniques for ecosystems are almost always destructive. This happens particularly as population densities increase, and as a result farming becomes more intensively practiced. This is because as demand for more land increases, the fallow period by necessity declines. The principal vulnerability is the nutrient-poor soil, pervasive in most tropical forests.

When biomass is extracted even for one harvest of wood or charcoal, the residual soil value is heavily diminished for further growth of any type of vegetation.

Kimpese, Democratic Republic of Congo.

Sometimes there are several cycles of slash-and-burn within a few years' time span. For example, in eastern Madagascar, the following scenario occurs commonly. The first wave might be cutting of all trees for wood use. A few years later, saplings are harvested to make charcoal, and within the next year the plot is burned to create a quick flush of nutrients for grass to feed the family zebu cattle. If adjacent plots are treated in a similar fashion, large-scale erosion will usually ensue, since there are no roots or temporary water storage in nearby canopies to arrest the surface runoff. Thus, any small remaining amounts of nutrients are washed away. The area is an example of desertification, and no further growth of any type may arise for generations.

The ecological ramifications of the above scenario are further magnified, because tropical forests are habitats for extremely biologically diverse ecosystems, typically containing large numbers of endemic and endangered species. Therefore, the role of slash-and-burn is significant in the current Holocene extinction.

Slash-and-char is an alternative that alleviates some of the negative ecological implications of traditional slash-and-burn techniques.

REFERENCES

1. Waters, Tony (2007). The Persistence of Subsistence Agriculture. Lanham: Lexington Books. p. 3. ISBN 978-0-7391-0768-3. OCLC 70334845.
2. Slash and burn. Encyclopedia of Earth.

3. Skegg, Martin (24 September 2011). "True Stories: Up in Smoke". The Guardian.
4. Elkan, Daniel (21 April 2004). "Slash-and-burn farming has become a major threat to the world's rainforest". The Guardian.
5. Jaime Awe, Maya Cities and Sacred Caves, Cu bola Books (2006)
6. Clark J.G.D. 1952, Farming: Clearance and Cultivation II Prehistoric Europe: The Economic Basis, Cambridge.
7. Semple E.C.1931, Ancient Mediterranean Forests and the Lumber Trade, vol. II., p. 261-296. Henry Holt et al., The Geography of the Mediterranean Region, New York.
8. Homer, e.g., Iliad XIII.11–13, Odyssey IX.22–24.
9. Darby, H.C., 1950, Domesday Woodland II Economic History Review, 2d ser.,III, London; Darby, H.C., 1956, The clearing of the Woodland in Europe II, p 186.
10. A W Liljenstrand wrote 1857 in his doctoral dissertation, "About Changing of Soil" (p. 5 ff.), that Tacitus discusses shifting cultivation: "arva per annos mutant".
11. Perkins and Marvin, Ex Editione Oberliniana, Harvard College Library, 1840 (Xxvi, 15–23).
12. Arenander E.O. 1923, Germanemas jordbrukskultur ornkring KristifØdelse // Berattelse over Det Nordiska Arkeologmotet i Stockholm 1922, Stockholm.
13. U.S. Department of Agriculture Lacey Act Guidance". USDA APHIS. October 26, 2011.
14. Stewart O.C. 1956, Fire as the First Great Force Employed by Man, II. Thomas W.L. Man's Role in Changing the Face of the Earth, Chicago.
15. Zwingle, E. (January 2005), "Italy before the Romans", National Geographic, Washington, D.C.
16. Late antique writers commonly used Getae for Goths mixing the peoples in the process.
17. G. Costa, 32. Also: De Rebus Geticis: O. Seyffert, 329; De Getarum (Gothorum) Origine et Rebus Gestis: W. Smith, vol 2 page 607
18. Alexander, Andrew C. Scott, David M.J.S. Bowman, William J. Bond, Stephen J. Pyne, Martin E. (2013). Fire on Earth : An Introduction. Hoboken: John Wiley & Sons Inc. p. 287. ISBN 1118570715.
19. Pyne, Stephen J. (1997). World Fire : The Culture of Fire on Earth (Pbk. ed.). Seattle: University of Washington Press. p. 85. ISBN 0295975938.
20. Svedjebruk.Hedmark 2010.Foreword by Ragnhild Queseth Haarstad Styreleder i Norsk Skogfinsk Museum. ISBN 978-82-93036-00-5.
21. "Slash and Burn Agriculture – An Overview of Slash and Burn". Geography. about.com. Retrieved 2013-06-22.
22. "Blog Archive " Farmer Power: The Continuing Confrontation between Subsistence Farmers and Development Bureaucrats". Ethnography.com. 2010-12-02. Retrieved 2013-06-22.

23. "Telkkämäki Nature Reserve". Outdoors.fi. 2013-05-14. Retrieved 2013-06-22.
24. "Jhum". banglapedia.org.
25. TI Trade (2011-01-17). "The Assam Tribune Online". Assamtribune.com. Retrieved 2013-06-22.
26. Tony Waters, The Persistence of Subsistence Agriculture. Lexington Books (2007), p. 48.
27. "The Forest Buddies". thinkquest.org. Retrieved 2015-12-16.

CHAPTER 6

Forest Ecology

Chapter Contents

- **Introduction**
- **Relationship to Other Branches of Ecology**
- **Community Diversity and Complexity**
- **Energy Flux**
- **Death And Regeneration**
- **Water**
- **Ecological Potential of Forestal Species**
- **Controlled Burn**
- **Ecological Thinning**
- **Reforestation**
- **Silviculture**
- **Tree Measurement**

Forest ecology is the science of how organisms react with each other, and interact with the environment of a forest ecosystem. This particular field of study is very important for those who are concerned about forest conservation. Without good data on the forest, what makes it healthy or unhealthy, there is no way a sustainable plan can be put into place.

INTRODUCTION

Forest Ecology is the scientific study of the interrelated patterns, processes, flora, fauna and ecosystems in forests. The management of forests is known as forestry, silviculture, and forest management. A forest ecosystem is a natural woodland unit consisting of all plants, animals and micro-organisms (Biotic components) in that area functioning together with all of the non-living physical (abiotic) factors of the environment. The forest ecosystem is very important.

The Daintree Rainforest in Queensland, Australia.

RELATIONSHIP TO OTHER BRANCHES OF ECOLOGY

Forest ecology is one branch of a biotically-oriented classification of types of ecological study (as opposed to a classification based on organizational level or complexity, for example population or community ecology). Thus, forests are studied at a number of organizational levels, from the individual organism to the ecosystem. However, as the term forest connotes an area inhabited by more than one organism, forest ecology most often concentrates on the level of the population, community or ecosystem. Logically, trees are an important component of forest research, but the wide variety of other life forms and abiotic components in most forests means that other elements, such as wildlife or soil nutrients, are often the focal point. Thus, forest ecology is a highly diverse and important branch of ecological study.

Forest ecology studies share characteristics and methodological approaches with other areas of terrestrial plant ecology. However, the presence of trees makes forest ecosystems and their study unique in numerous ways.

COMMUNITY DIVERSITY AND COMPLEXITY

Since trees can grow larger than other plant life-forms, there is the potential for a wide variety of forest structures (or physiognomies). The infinite number of possible spatial arrangements of trees of varying size and species makes for a highly intricate and diverse micro-environment in which environmental variables such as solar radiation, temperature, relative humidity, and wind speed can vary considerably over large and small distances. In addition, an important proportion of a forest ecosystem's biomass is often underground, where soil structure, water quality and quantity, and levels of various soil nutrients can vary greatly. Thus, forests are often highly heterogeneous environments compared to other terrestrial plant communities. This heterogeneity in turn can enable great biodiversity of species of both plants and animals. It also affects the design of forest inventory sampling strategies, the results of which are sometimes used in ecological studies. A number of factors within the forest affect biodiversity; primary factors enhancing wildlife abundance and biodiversity are the presence of diverse tree species within the forest and the absence of even aged timber management. For example, the wild turkey thrives when uneven heights and canopy variations exist and its numbers are diminished by even aged timber management.

ENERGY FLUX

Forest ecologists are interested in the effects of large disturbances, such as wildfires. Montana, United States.

Forests accumulate large amounts of standing biomass, and many are capable of accumulating it at high rates, i.e. they are highly productive. Such high levels of biomass and tall vertical structures represent large stores of potential energy that can be converted to kinetic energy under the right circumstances. Two such conversions of great importance are fires and treefalls, both of which radically alter the biota and the physical environment where they occur. Also, in forests of high productivity, the rapid growth of the trees themselves induces biotic and environmental changes, although at a slower rate and lower intensity than relatively instantaneous disturbances such as fires.

DEATH AND REGENERATION

Woody material, often referred to as coarse woody debris, decays relatively slowly in many forests in comparison to most other organic materials, due to a combination of environmental factors and wood chemistry (see lignin). Trees growing in arid and/or cold environments do so especially slowly. Thus, tree trunks and branches can remain on the forest floor for long periods, affecting such things as wildlife habitat, fire behavior, and tree regeneration processes.

WATER

Lastly, forest trees store large amounts of water because of their large size and anatomical/ physiological characteristics. They are therefore important regulators of hydrological processes, especially those involving groundwater hydrology and local evaporation and rainfall/snowfall patterns. Thus, forest ecological studies are sometimes closely aligned with meteorological and hydrological studies in regional ecosystem or resource planning studies. Perhaps more importantly the duff or leaf litter can form a major repository of water storage. When this litter is removed or compacted (e.g. through grazing or human overuse), erosion and flooding are exacerbated as well as deprivation of dry season water for forest organisms.

ECOLOGICAL POTENTIAL OF FORESTAL SPECIES

The ecological potential of a particular species is a measure of its capacity to effectively compete in a given geographical area, ahead of other species, as they all try to occupy a natural space. For some areas it has been quantified, as for instance by Hans-Jürgen Otto, for central Europe. He takes three groups of parameters:

- Related to site requirements: Tolerance to low temperatures, tolerance to dry climate, frugality.
- Specific qualities: Shade tolerance, height growth, stability, longevity, regeneration capacity.
- Specific risks: Resistance to late freezing, resistance to wind/ice storm, resistance to fire, resistance to biotic agents.

Every parameter is scored between 0 and 5 for each considered species, and then a global mean value calculated. A value above 3.5 is considered high, below 3.0 low, and intermediate for those in between. In this study *Fagus sylvatica* has a score of 3.82, *Fraxinus excelsior* 3.08 and *Juglans regia* 2.92; and are examples of the three categories.

CONTROLLED BURN

Controlled or **prescribed burning**, also known as **hazard reduction burning**(HRB), **backfire**, or **swailing**, is a technique sometimes used in forest management, farming, prairie restoration or greenhouse gas abatement. Fire is a natural part of both forest and grassland ecology and controlled fire can be a tool for foresters. Hazard reduction or controlled burning is conducted during the cooler months to reduce fuel buildup and decrease the likelihood of serious hotter fires. Controlled burning stimulates the germination of some desirable forest trees, and reveals soil mineral layers which increases seedling vitality, thus renewing the forest. Some cones, such as those of Lodgepole Pine and Sequoia, are serotinous, as well as many chaparral shrubs, meaning they require heat from fire to open cones to disperse seeds.

An experimental burn in Canada; note the measuring equipment in front.

Firing the woods in a South Carolina forest with a custom made driptorch mounted on an ATV. The device spits flaming fuel oil from the side, instantly igniting the leaf litter.

A prescribed burn in a *Pinus nigra* stand in Portugal.

Controlled burning of a field outside of Statesboro, Georgia, United States in preparation for spring planting.

An aerial view of a controlled burn in Helderberg Nature Reserve in South Africa bordering the city of Cape Town. In South Africa controlled burns are important for maintaining the ecological health of indigenous fynbos as well as reducing the intensity of future burns.

Japanese controlled burn (Hokkaido).

In industrialized countries, controlled burning is usually overseen by fire control authorities for regulations and permits. The party responsible must delineate the intended time and place. Obtaining a permit may not limit liability if the fire burns out of control.

There are two basic causes of wildfires. One is natural (lightning) and the other is people. Controlled burns have a long history in wildland management. Pre-agricultural societies used fire to regulate both plant and animal life. Fire history studies have documented periodic wildland fires ignited by indigenous peoples in North America and Australia.

Fires, both naturally caused and prescribed, were once part of natural landscapes in many areas. Studies have shown that between the mid Holocene and the 17th century AD, wildland fires annually burned between 45.0% and 87.5% of present-day California's total land, for example. These practices ended in the early 20th century when US fire policies were enacted with the goals of suppressing all fires. Since 1995, the US Forest Service has slowly incorporated burning practices into its forest management policies.

Back Burning

Back burning is a technique utilized in controlled burning and during wildfire events. While controlled burns utilize back burning during planned fire events to create a "black line", back burning or backfiring is also done to stop a wildfire that is already in progress. Back burning is a way of reducing the amount of flammable material during a controlled burn or wildfire by starting small fires along a man made or natural firebreak in front of a main fire front. It is called back burning because the small fires are designed to 'burn back towards the main fire front' and are usually burning and traveling against ground level winds. Back burning reduces the amount of fuel that's available to the main fire by the time that it reaches the burnt area. Firebreaks are often used as an anchor point to start a line of fires along natural or manmade features such as a river, road or a bulldozed clearing.

Forest Use

Another consideration is the issue of fire prevention. In Florida, during the drought in 1995, catastrophic wildfires burned numerous homes. But forestry managers in the Florida Division of Forestry noted that the underlying problem was previous cessation of controlled burning, resulting from complaints by homeowners. Each year additional leaf litter and dropped branches increased the likelihood of a hot and uncontrollable fire.

Controlled burns are sometimes ignited using a tool known as the driptorch, which allows a steady stream of flaming fuel to be directed to the ground as needed. Variations on the driptorch can be used such as the *helitorch*, which is mounted on a helicopter, or other improvised devices such as mounting a driptorch-like device on the side of a vehicle. A pyrotechnic device known as a fusee can be used for ignition in nearby fuels while a Very pistol can be for fuels farther away.

For the burning of slash, waste materials left over from logging, there are several types of controlled burns. Broadcast burning is the burning of scattered slash over a wide area. Pile burning is gathering up the slash into piles before burning. These burning piles may be referred to as bonfires. High temperatures can harm the soil, damaging it physically, chemically or sterilizing it. Broadcast burns tend to have lower temperatures and will not harm the soil as much as pile burning, though steps can be taken to treat the soil after a burn. In lop and scatter burning, slash is left to compact over time, or is compacted with machinery. This produces a lower intensity fire, as long as the slash is not packed too tightly. However, soil may be damaged if machinery is used to compress the slash.

Controlled burning reduces fuels, may improve wildlife habitat, controls competing vegetation, improves short term forage for grazing, improves accessibility, helps control tree disease, and perpetuates fire dependent species. In mature longleaf pine forest, it helps maintain habitat for endangered red-cockaded woodpeckers in their sandhill and flatwoods habitats.Fire is also felt to be a crucial element of the recovery of the threatened Louisiana pine snake in the longleaf pine forests of central Louisiana and eastern Texas. However, many scientists disagree with such a simplistic approach, and indicate that each forest must be assessed on its own merit.

In the wild, many trees depend on fire as a successful way to clear out competition and release their seeds. In particular, the Giant sequoia depends on fire to reproduce: the cones of the tree open after a fire releases their seeds, the fire having cleared all competing vegetation. Due to fire suppression efforts during the early and mid 20th century, low-intensity fires no longer occurred naturally in many groves, and still do not occur in some groves today. The suppression of fires also led to ground fuel build-up and the dense growth which posed the risk of catastrophic wildfires. It wasn't until the 1970s that the National Park Service began systematic fires for the purpose of new seed growth. Eucalyptus regnans or Mountain Ash of Australia also depends on fire but in a different fashion. They carry their seeds in capsules which can deposit at any time of the year. Being flammable, during a fire the capsules drop nearly all of them and the fire consumes the Eucalypt adults, but most of the seeds survive using the ash as a source of nutrients; at their rate of growth, they quickly dominate the land and a new Eucalyptus forest grows.

Agricultural Use

In addition to forest management, controlled burning is also used in agriculture. In the developing world, this is often referred to as slash and burn. In industrialized nations, it is seen as one component of shifting cultivation, as a part of field preparation for planting. Often called **field burning**, this technique is used to clear the land of any existing crop residue as well as kill weeds and weed seeds. Field burning is less expensive than most other methods such as herbicides or tillage, but because it does produce smoke and other fire-related pollutants, its use is not popular in agricultural areas bounded by residential housing.

In the United States, field burning is a legislative and regulatory issue at both the federal and state levels of government.

Controversy

In Oregon, field burning has been widely used by grass seed farmers as a method for clearing fields for the next round of planting, as well as revitalizing serotinous grasses that require fire in order to grow seed again. The Oregon Department of Environmental Quality began requiring a permit for farmers to burn their fields in 1981, but the requirements became stricter in 1988 following a multi-car collision in which smoke from field burning near Albany, Oregon, obscured the vision of drivers on Interstate 5, leading to a 23-car collision in which 5 people died and 37 were injured. This resulted in more scrutiny of field burning and proposals to ban field burning in the state altogether.

In the European Union burning crop stubble after harvest is used by farmers for plant health reasons under several restrictions in the cross compliance regulations.

Heathland Use

The controlled burning or swailing of heathland is used in the United Kingdom and other countries as a conservation tool. In Scotland it is known as "muirburn". Often it is used as a tool for creating fire breaks to reduce the risk of dangerous outbreaks but is also an important mechanism for preventing succession to woodier vegetation and plays an important role in the life cycle of heather species. Red grouse eat heather shoots and as grouse shooting has long been an expensive sport the landowners need to ensure a good crop of grouse so each year parts of the moor are burned of old heather to cause fresh shoots.

Greenhouse gas abatement

Northern California fire crews start a backfire to stop the Poomacha fire from advancing westward.

Controlled burns on Australian savannas can result in an overall reduction of greenhouse gas emissions. One working example is the West Arnhem Fire Management Agreement, started to bring "strategic fire management across 28,000 square kilometres (11,000 sq mi) of Western Arnhem Land" to partially offset greenhouse gas emissions from a liquefied natural gas plant in Darwin, Australia. Deliberately starting controlled burns early in the dry season results in a mosaic of burnt and unburnt country which reduces the area of stronger, late dry season fires.Also known as "patch burning". To minimise the impact of smoke, burning should be restricted to daylight hours whenever possible.

ECOLOGICAL THINNING

Fresh snow partially covers epixylic Rough-stalked Feather-moss *(Brachythecium rutabulum)*, growing on a dead hybrid black poplar *(Populus x canadensis)*, thinned in 2008 to create a waterside path. The last stage of the moss lifecycle is shown, where the sporophytes are visible before dispersion of their spores: the calyptra (**1**) is still attached to the sporangium (**2**). The tops of the gametophytes (**3**) can be discerned as well. Inset shows the surrounding, black poplars growing on sandy loam on the bank of a kolk, with the detail area marked.

Ecological thinning is a silvicultural technique used in forest management that involves cutting trees to improve functions of a forest other than timber production.

Although thinning originated as a man-made forest management tool, aimed at increasing timber yields, the shift from production forests to multifunctional forests brought with it the cutting of trees to manipulate an ecosystem for various reasons, ranging from removing non-native species from a plot to removing poplars growing on a riverside beach aimed at recreational use.

Since the 1970s, leaving the thinned trees on the forest floor has become an increasingly common policy: wood can be decomposed in a more natural fashion, playing an important role in increasing biodiversity by providing habitat to various invertebrates, birds and small mammals. Many fungi (e.g. *Calocera viscosa*) and mosses are saproxylic or epixylic as well (e.g. *Marchantiophyta*) – some moss species completing their entire life-cycle on a single log.

Where trees are managed under a commercial regime, competition is reduced by removing adjacent stems that exhibit less favourable timber quality potential. When left in a natural state trees will "self-thin", but this process can be unreliable in some circumstances. Examples of this can be found in the Buxus - Ironbark forests and woodlands of Victoria (Australia) where a large proportion of trees are coppice, resultant from timber cutting in decades gone by.

Ecophysiological Repercussions of Thinning

Thinning decreases canopy closure and increases the penetration of solar radiation into the canopy. The photosynthetic efficiency of this energy is improved, and needle retention is prolonged, especially in the lower parts of the crown. The root system, crown length, crown diameter, and crown area all increase after thinning. Even if soil evaporation and individual tree transpiration increases after thinning, total evapo-transpiration at stand level tends to decrease; canopy water interception is reduced and throughfall increased, so that tree-water status usually improves after thinning.

Radial annual growth is an integrative index of tree physiological response to environmental variation. Working with Norway spruce, which can be expected to behave in some respects similarly to white spruce, Misson et al. showed that a reduction in stand density alters the classical climate–growth relationship. At individual tree level, thinning could be used to increase tree resistance to drought stress. Nevertheless, this effect is limited when site conditions are limiting. Misson et al. concluded that heavy thinning should be applied when forest decline is expected from drought stress. Furthermore, stands on dry sites should be thinned more heavily because such sites cannot support high-density stands.

Aussenac, also working with Norway spruce, investigated the response to thinning. Thinning altered the temporal evolution of radial growth at all frequencies in radial growth chronologies. Previous studies had shown, not surprisingly, that thinning decreases the canopy closure and encourages solar radiation to penetrate the canopy. Solar energy then becomes more important inside the crown. The improved photosynthetic efficiency of this energy influences the retention time of needles, especially in the lower part of the crown. Furthermore, the length, diameter, and area, of the crown, and the size of the root system all increase after thinning.

These modifications at the crown level influence photosynthate production positively as long as water supply is not limiting. Even if soil evaporation and individual tree transpiration are more important after thinning, total evapo-transpiration at stand level tends to decrease. Furthermore, with the reduced importance of leaf area index (LAI), thinning reduces canopy water interception and increases throughfall. This explains why the soil moisture content increases in a thinned stand.Thus, during the vegetation period, tree-water status is usually better in a thinned stand than in a high-density stand.

For semi-tolerant species such as Norway spruce and white spruce, solar energy is less limiting than for intolerant species. This can account for the exponential relationship found by Misson et al. between thinning intensity and mean radial growth. Only when thinning was relatively severe did important radial growth variation occur. Furthermore, the ecophysiological advantage of an improved water supply could be counterbalanced by limiting site conditions.

For Norway spruce in the Belgian Ardennes, Misson et al. recommended that stand basal area should not exceed 26 m^2/ha on a dry site, or 29 m^2/ha on a moist site if the objective is to maintain high long-term radial growth.

Misson et al. found that adaptations to thinning cover a continuous range of different time scales, with long-, medium-, and short-term growth variations being controlled by very different internal factors influenced by the environment. Misson et al.ascribed what they called long- and medium-term radial growth variation of individual trees mainly to structural adaptation, such as enlargement of the crown or the root system. Short-term radial growth variations, on the other hand, were considered to be due mainly to physiological acclimatisation, with response to factors such as stomatal conductance regulation, variation of photosynthetic capacity, and respiration. Nevertheless, it is clear that the adaptations are linked, and that radial growth is the integrative response of this mutual inter-relationship.

Ecological Thinning Research

Research programs under way in various parts of the world (e.g. United States and Australia) are aimed at providing an alternate approach in forest management where conservation objectives are a high priority. Methods of ecological thinning being developed on silvicultural techniques for local forest types. Ecological thinning is being developed using two principles: 1. appropriate stem reduction to reduce competition and 2. retention of trees (selection) that are more suitable for wildlife (i.e. not timber production). An example of ecological thinning research is the project in Victoria's Box-Ironbark forests, investigating various thinning and timber removal methods under an Adaptive Management or AEM framework. The primary objective is to generate (over time) a number of forest habitat values (i.e. tree hollows) that are crucial for wildlife conservation.

REFORESTATION

Reforestation is the natural or intentional restocking of existing forests and woodlands that have been depleted, usually through deforestation. Reforestation can be used to rectify or improve the quality of human life by soaking up pollution and dust from the air, rebuild natural habitats and ecosystems, mitigate global warming since forests facilitate biosequestration of atmospheric carbon dioxide, and harvest for resources, particularly timber, but also non-timber forest products.

Tropical tree nursery at Planeta Verde Reforestación S.A.'s plantation in Vichada Department, Colombia

A 15-year-old reforested plot of land

A 21-year-old plantation of red pine in Southern Ontario

The term reforestation is similar to afforestation, the process of restoring and recreating areas of woodlands or forests that may have existed long ago but were deforested or otherwise removed at some point in the past. Sometimes the term *re-afforestation* is used to distinguish between the original forest cover and the later re-growth of forest to an area. **Forestation** is the establishment of forest growth on areas that either had forest or lacked it.

Special tools, e.g. tree planting bars, are used to make planting of trees easier and faster.

Management

A debated issue in managed reforestation is whether or not the succeeding forest will have the same biodiversity as the original forest. If the forest is replaced with only one species of tree and all other vegetation is prevented from growing back, a monoculture forest similar to agricultural crops would be the result. However, most reforestation involves the planting of different seedlots of seedlings taken from the area, often of multiple species. Another important factor is the natural regeneration of a wide variety of plant and animal species that can occur on a clear cut. In some areas the suppression of forest fires for hundreds of years has resulted in large single aged and single species forest stands. The logging of small clear cuts and/or prescribed burning, actually increases the biodiversity in these areas by creating a greater variety of tree stand ages and species.

For Harvesting

Reforestation need not be only used for recovery of accidentally destroyed forests. In some countries, such as Finland, many of the forests are *managed* by the wood products and pulp and paper industry. In such an arrangement, like other crops, trees are planted to replace those that have been cut. In such circumstances, the industry can cut the trees in a way to allow easier reforestation. The wood products industry systematically replaces many of the trees it cuts, employing large numbers of summer workers for tree planting work. For example, in 2010, Weyerhaeuser reported planting 50 million seedlings.However replanting an old-growth forest with a plantation is not replacing the old with the same characteristics in the new.

In just 20 years, a teak plantation in Costa Rica can produce up to about 400 m^3 of wood per hectare. As the natural teak forests of Asia become more scarce or difficult to obtain, the prices commanded by plantation-grown teak grows higher every year. Other species such as mahogany grow more slowly than teak in Tropical America but are also extremely valuable. Faster growers include pine, eucalyptus, and *Gmelina*.

Reforestation, if several indigenous species are used, can provide other benefits in addition to financial returns, including restoration of the soil, rejuvenation of local flora and fauna, and the capturing and sequestering of 38 tons of carbon dioxide per hectare per year.

The reestablishment of forests is not just simple tree planting. Forests are made up of a community of species and they build dead organic matter into soils over time. A major tree-planting program could enhance the local climate and reduce the demands of burning large amounts of fossil fuels for cooling in the summer.

For Climate Change Mitigation

Forests are an important part of the global carbon cycle because trees and plants absorb carbon dioxide through photosynthesis. By removing this greenhouse gas from the air, forests function as terrestrial carbon sinks, meaning they store large amounts of carbon. At any time, forests account for as much as double the amount of carbon in the atmosphere. Even as more anthropogenic carbon is produced, forests remove around three billion tons of anthropogenic carbon every year. This amounts to about 30% of all carbon dioxide emissions from fossil fuels. Therefore, an increase in the overall forest cover around the world would tend to mitigate global warming.

There are four major strategies available to mitigate carbon emissions through forestry activities: increase the amount of forested land through a reforestation process; increase the carbon density of existing forests at a stand and landscape scale; expand the use of forest products that will sustainably replace fossil-fuel emissions; and reduce carbon emissions that are caused from deforestation and degradation.

Achieving the first strategy would require enormous and wide-ranging efforts. However, there are many organizations around the world that encourage tree-planting as a way to offset carbon emissions for the express purpose of fighting climate change. For example, in China, the Jane Goodall Institute, through their Shanghai Roots & Shoots division, launched the Million Tree Project in Kulun Qi, Inner Mongolia to plant one million trees to stop desertification and help curb climate change. China has used 24 billion metres squared of new forest plantation and natural forest regrowth to offset 21% of Chinese fossil fuel emissions in 2000.[:1456] In Java, Indonesia each newlywed couple is to give whoever is sermonizing their wedding 5 seedlings to combat global warming. Each couple that wishes to have a divorce has to give 25 seedlings to whoever divorces them.

The second strategy has to do with selecting species for tree-planting. In theory, planting any kind of tree to produce more forest cover would absorb more carbon dioxide from the atmosphere. On the other hand, a genetically modified tree specimen might grow much faster than any other regular tree. Some of these trees are already being developed in the lumber and biofuel industries. These fast-growing trees would not only be planted for those industries but they can also be planted to help absorb carbon dioxide faster than slow-growing trees.

Extensive forest resources placed anywhere in the world will not always have the same impact. For example, large reforestation programs in boreal or subarctic regions have a limited impact on climate mitigation. This is because it substitutes a bright snow-dominated region that reflects the sunlight with dark forest canopies. A study from the National Center for Atmospheric Research in Boulder, Colorado, USA, found that trees in temperate latitudes have a net warming effect on the atmosphere. The heat that dark leaves release without absorbing outweighs the carbon they sequester. On the other hand, a positive example would be reforestation projects in tropical regions, which would lead to a positive biophysical change such as the formation of clouds. These clouds would then reflect the sunlight, creating a positive impact on climate mitigation.[:1457]

There is an advantage to planting trees in tropical climates with wet seasons. In such a setting, trees have a quicker growth rate because they can grow year-round. Trees in tropical climates have, on average, larger, brighter, and more abundant leaves than non-tropical climates. A study of the girth of 70,000 trees across Africa has shown that tropical forests are soaking up more carbon dioxide pollution than previously realized. The research suggests almost one fifth of fossil fuel emissions are absorbed by forests across Africa, Amazonia and Asia. Simon Lewis, a climate expert at the University of Leeds, who led the study, said: "Tropical forest trees are absorbing about 18% of the carbon dioxide added to the atmosphere each year from burning fossil fuels, substantially buffering the rate of change."

It is also important to deal with the rate of deforestation. At this point, there are 13 billion metres squared of tropical regions that are deforested every year. There is potential for these regions to reduce rates of deforestation by 50% by 2050, which would be a huge contribution to stabilize the global climate.[:1456]

On Abandoned Farmland

With increased intensive agriculture and urbanization, there is an increase in the amount of abandoned farmland. By some estimates, for every half a hectare of original old-growth forest cut down, more than 20 hectares of new secondary forests are growing, even though they do not have the same biodiversity as the original forests and original forests store 60% more carbon than these new secondary forests. According to a study in Science, promoting regrowth on abandoned farmland could offset years of carbon emissions.

Promotion Strategies

Land Rights

Transferring land rights to indigenous inhabitants is argued to efficiently conserve forests.

Transferring rights over land from public domain to its indigenous inhabitants is argued to be a cost effective strategy to conserve forests. This includes the protection of such rights entitled in existing laws, such as India's Forest Rights Act. The transferring of such rights in China, perhaps the largest land reform in modern times, has been argued to have increased forest cover. In Brazil, forested areas given tenure to indigenous groups have even lower rates of clearing than national parks.

Incentives

Some incentives for reforestation can be as simple as a financial compensation. Streck and Scholz (2006) explain how a group of scientists from various institutions have developed a compensated reduction of deforestation approach which would reward developing countries that disrupt any further act of deforestation. Countries that participate and take the option to reduce their emissions from deforestation during a committed period of time would receive financial compensation for the carbon dioxide emissions that they avoided.[:875] To raise the payments, the host country would issue government bonds or negotiate some kind of loan with a financial institution that would want to take part in the compensation promised to the other country. The funds received by the country could be invested to help find alternatives to the extensive cutdown of forests. This whole process of cutting emissions would be voluntary, but once the country has agreed to lower their emissions they would be obligated to reduce their emissions. However, if a country was not able to meet their obligation, their target would get added to their next commitment period. The authors of these proposals see this as a solely government-to-government agreement; private entities would not participate in the compensation trades.[:876]

Examples

Forest regrowth in Mount Baker-Snoqualmie National Forest, Washington state, USA.

Sub-Saharan Africa

Plans to plant a nine-mile width of trees on the Southern Border of the Sahara Desert. The Great Green Wall initiative is a pan-African proposal to "green" the continent from west to east in order to battle desertification. It aims at tackling poverty and the degradation of soils in the Sahel-Saharan region, focusing on a strip of land of 15 km (9 mi) wide and 7,500 km (4,750 mi) long from Dakar to Djibouti.

Canada

In Canada, overall forest cover is increasing over the last decades.

China

In China, extensive replanting programs have existed since the 1970s. Programs have had overall success. The forest cover has increased from 12% of China's land area to 16%. However, specific programs have had limited success. The "Green Wall of China", an attempt to limit the expansion of the Gobi Desert is planned to be 2,800 miles (4,500 km) long and to be completed in 2050. China plans to plant 26 billion trees in the next decade that is two trees for every Chinese citizen per year. China requires that students older than 11 years old plant one tree a year until their high school graduation.

Germany

Reforestation is required as part of the federal forest law. 31% of Germany is forested, according to the second forest inventory of 2001–2003. The size of the forest area in Germany increased between the first and the second forest inventory due to forestation of degenerated bogs and agricultural areas.

India

Jadav Payeng had received national awards for reforestation efforts, known as the "Molai forest".

Japan

Mass environmental and human-body pollution along with relating deforestation, water pollution, smoke damage, and loss of soils caused by mining operations in Ashio, Tochigi became the first environmental social issue in Japan, efforts by Shōzō Tanaka had grown to large campaigns against copper operation. This led to the creation of 'Watarase Yusuichi Pond', to settle the pollution which is a Ramsar site today. Reforestation was conducted as a part of afforestation due to inabilities of self-recovering by the natural land itself due to serious soil pollution and loss of woods consequence in loss of soils for plants to grow, thus needing artificial efforts involving introducing of healthy soils from outside. Starting since in 1897, about 50% of once bald mountains backed to green.

Lebanon

For thousands of years Lebanon was covered by forests, one particular species of interest, *Cedrus libani* or Lebanon Cedar was exceptionally valuable timber species. Virtually every ancient culture that shared the Mediterranean Sea harvested these trees, the Ancient Egyptians, the Greeks, the Roman Empire, early Christians, the Muslims, Persians, Assyrians and Babylonians. Extensive deforestation has occurred, with only small remnants of the original forests surviving. Deforestation has been particularly severe in Lebanon and on Cyprus. The Lebanon Reforestation Initiative aims to restore Lebanon's native forests. Projects financed locally and by international charity are performing extensive reforestation of cedar being carried out in the Mediterranean region, particularly in Lebanon and Turkey, where over 50 million young cedars are being planted annually.

Philippines

The Philippines increased its forest area to 240,000 hectares (593,053 acre) per year from 2010 to 2015. In 2011-2015 period, reforestation through the government's National Greening Program covered 1,258,692 hectares of land, which is 105 percent more than the 1,200,000 hectares planting target for the same period. An estimated eight million hectares of land nationwide need reforestation and rehabilitation, according to the Forest Management Bureau.

United States

It is the stated goal of the US Forest Service to manage forest resources sustainably. This includes reforestation after timber harvest, among other programs.

Organizations

Trees for the Future has assisted more than 170,000 families, in 6,800 villages of Asia, Africa and the Americas, to plant over 35 million trees.

Wangari Maathai, 2004 Nobel Peace Prize recipient, founded the Green Belt Movement which planted over 47 million trees to restore the Kenyan environment.

Shanghai Roots & Shoots, a division of the Jane Goodall Institute, launched The Million Tree Project in Kulun Qi, Inner Mongolia to plant one million trees to stop desertification and alleviate global warming.

Criticisms

Reforestation competes with other land uses such as food production, livestock grazing and living space for further economic growth. Reforestation often has the tendency to create large fuel loads, resulting in significantly hotter combustion than fires involving low brush or grasses. Reforestation can divert large amounts of water from other activities. Reforesting sometimes results in extensive canopy creation that prevents growth of diverse vegetation in the shadowed areas and generating soil conditions that hamper other types of vegetation. Trees used in some reforesting efforts (e.g., *Eucalyptus globulus*) tend to extract large amounts of moisture from the soil, preventing the growth of other plants.

There is also the risk that through a forest fire or insect outbreak much of the stored carbon in a reforested area could make its way back to the atmosphere.[:1456] Reduced harvesting rates and fire suppression have caused an increase in the forest biomass in the western United States over the past century. This causes an increase of about a factor of four in the frequency of fires due to longer and hotter dry seasons.[:1456]

SILVICULTURE

Silviculture is the practice of controlling the establishment, growth, composition, health, and quality of forests to meet diverse needs and values.

The name comes from the Latin *silvi-* (forest) + culture (as in growing). The study of forests and woods is termed silvology. Silviculture also focuses on making sure that the treatment(s) of forest stands are used to preserve and to better their productivity.

Generally, silviculture is the science and art of growing and tending forest crops, based on a knowledge of silvics, i.e., the study of the life history and general characteristics of forest trees and stands, with particular reference to locality factors.More particularly, silviculture is the theory and practice of controlling the establishment, composition, constitution, and growth of forests. No matter how forestry as a science is constituted, the kernel of the business of forestry is silviculture, as it includes direct action in the forest, and in it all economic objectives and technical considerations ultimately converge. The kernel of silviculture is regeneration.

Suggestions for how best to go about the job, presented by Jeglum et al. (2003), though aimed primarily at the boreal forest in Ontario, merit wider consideration. The 110-page publication describes Best Management Practices, first by general principles, then by sensitive sites. Illustrations are plentiful and are well chosen to complement this excellent text.

To some the distinction between forestry and silviculture is that silviculture is applied at the stand level and forestry is broader. For example, John D. Matthews says "complete regimes for regenerating, tending, and harvesting forests" are called "silvicultural systems".

So adaptive management is required for silviculture, whereas forestry can be natural, conserved land without a stand level treatment being applied. A common taxonomy divides silviculture into regenerating, tending and harvesting techniques.

An employee of the White Mountain Apache Tribe forestry department uses a hoedad to dig a hole for a ponderosa pine seedling in the shade of a burned tree stump, which protects the seedling from windy conditions that dry the trees

Silvicultural Systems

The origin of forestry in German speaking Europe has defined silvicultural systems broadly as **Hochwald** - High forest, **Mittelwald** - Coppice with standards (and Compound coppice), Short rotation coppice and **Niederwald** - Coppice. There are other systems too. These varied silvicultural systems include several **harvesting** methods, which are often wrongly said to be a silvicultural systems, but may also be called as rejuvenating or regenerating method depending on the purpose.

Hochwald is further subdivided in German into: **Altersklassenwald** - Age class forest which includes:

- even aged forestry
    - Kahlschlag - Clear cutting
    - Schirmschlag - Shelterwood cutting
    - Seed-tree method
- un-even aged forestry
- Femelschlag (Femelschlagverfahren) - The Femel selection cutting (Group selection cutting),
- Saumschlag (Saumfemelschlagbetrieb) - Strip selection cutting (strip-and-group felling system)
- Schirmkeilschlag - Shelterwood wedge cutting
- Mischformen - mixed form regeneration methods.

and **Dauerwald** - Continuous cover forestry, in which are found:

- un-even aged forestry

- Plenterwald - Selection forest
- Zielstärkennutzung - Target diameter harvesting.

From these names the impression is given that these are neatly defined systems, but in practice there are a variety of variations within these harvesting methods according to local ecology and site conditions. While an individual location can be identified where it could be said to be the archetypal form of harvesting technique (they all originated somewhere with a particular forester, and have been described in the scientific literature), and broad generalizations can be made these are merely rules of thumb rather than strict blue prints on how techniques might be applied. This misunderstanding has meant that many older English textbooks did not capture the true complexity of silviculture as practiced where it originated in Mitteleuropa.

This silviculture was culturally predicated on wood production in temperate and boreal climates and did not deal with tropical forestry. The misapplication of this philosophy to those tropical forests has been problematic. There is also an alternative silvicultural tradition which developed in Japan and thus created a different biocultural landscape called satoyama.

After harvesting comes **regeneration**, which may be split into Natural and Artificial (see below). And **tending** which includes release treatments, pruning, thinning and intermediate treatments. As it can be imagined it maybe that any of these 3 phases (harvesting - regeneration - tending) may happen at the same time, within a stand, depending on the goal for that particular stand.

Regeneration

Regeneration is basic to the continuation of forested, as well as to the afforestation of treeless land. Regeneration can take place through self-sown seed ("natural regeneration"), by artificially sown seed, or by planted seedlings. In whichever case, the performance of regeneration depends on its growth potential and the degree to which its environment allows the potential to be expressed. Seed, of course, is needed for all regeneration modes, both for natural or artificial sowing and for raising planting stock in a nursery.

Natural Regeneration is

Human-assisted natural regeneration means establishment of a forest age class from natural seeding or sprouting in an area after harvesting in that area through selection cutting, shelter (or seed-tree) harvest, soil preparation, or restricting the size of a clear-cut stand to secure natural regeneration from the surrounding trees.

The process of natural regeneration involves the renewal of forests by means of self-sown seeds, root suckers, or coppicing. In natural forests, conifers rely almost entirely on regeneration through seed. Most of the broadleaves, however, are able to regenerate by the means of emergence of shoots from stumps (coppice) and broken stems.

Seedbed Requirements

Any seed, self-sown or artificially applied, requires a seedbed suitable for securing germination.

In order to germinate, a seed requires suitable conditions of temperature, moisture, and aeration. For seeds of many species, light is also necessary, and facilitates the germination of seeds in other species, but spruces are not exacting in their light requirements, and will germinate without light. White spruce seed germinated at 35 °F (1.7 °C) and 40 °F (4.4 °C) after continuous stratification for one year or longer and developed radicles < 6 cm long in the cold room. When exposed to light, those germinants developed chlorophyll and were normally phototropic with continued elongation.

For survival in the short and medium terms, a germinant needs: a continuing supply of moisture; freedom from lethal temperature; enough light to generate sufficient photosynthate to support respiration and growth, but not enough to generate lethal stress in the seedling; freedom from browsers, tramplers, and pathogens; and a stable root system. Shade is very important to the survival of young seedlings. In the longer term, there must be an adequate supply of essential nutrients and an absence of smothering.

In undisturbed forest, decayed windfallen stemwood provides the most favourable seedbed for germination and survival, moisture supply being dependable, and the elevation of seedlings somewhat above the general level of the forest floor reduces the danger of smothering by leaves and snow-pressed minor vegetation; nor is such a microsite likely to be subject to flooding. Advantages conferred by those microsites include: more light, higher temperatures in the rooting zone, and better mycorrhizal development. In a survey in the Porcupine Hills, Manitoba, 90% of all spruce seedlings were rooted in rotten wood.

Mineral soil seedbeds are more receptive than the undisturbed forest floor, and are generally moister and more readily rewetted than the organic forest floor. However, exposed mineral soil, much more so than organic-surfaced soil, is subject to frost heaving and shrinkage during drought. The forces generated in soil by frost or drought are quite enough to break roots.

The range of microsites occurring on the forest floor can be broadened, and their frequency and distribution influenced by site preparation. Each microsite has its own microclimate. Microclimates near the ground are better characterized by vapour pressure deficit and net incident radiation, rather than the standard measurements of air temperature, precipitation, and wind pattern.

Aspect is an important component of microclimate, especially in relation to temperature and moisture regimes. Germination and seedling establishment of Engelmann spruce were much better on north than on south aspect seedbeds in the Fraser Experimental Forest, Colorado; the ratios of seeds to 5-year-old seedlings were determined as 32:1, 76:1, and 72:1 on north aspect bladed-shaded, bladed-unshaded, and undisturbed-shaded seedbeds, respectively. Clearcut openings of 1.2 ha to 2.0 ha adjacent to an adequate seed source, and not more than 6 tree heights wide, could be expected to secure acceptable regeneration (4,900, 5-year-old trees/ha), whereas on undisturbed-unshaded north aspects, and on all seedbed treatments tested on south aspects, seed to seedling ratios were so high that the restocking of any clearcut opening would be questionable.

At least 7 variable factors may influence seed germination: (1) seed characteristics, (2) light, (3) oxygen, (4) soil reaction (ph), (5) temperature, (6) moisture, and (7) seed enemies. Moisture and temperature are the most influential, and both are affected by exposure. The difficulty of securing natural regeneration of Norway spruce and Scots pine in northern Europe led to the adoption of various forms of reproduction cuttings that provided partial shade or protection to seedlings from hot sun and wind. The main objective of echeloned strips or border-cuttings with northeast exposure was to protect regeneration from overheating, and was originated in Germany and deployed successfully by Alarik (1925) and others in Sweden. On south and west exposures, direct insolation and heat reflected from tree trunks often result in temperatures lethal to young seedlings, as well as desiccation of the surface soil, which inhibits germination. The sun is less injurious on eastern exposures because of the lower temperature in the early morning, related to higher humidity and presence of dew.

Baldwin (1933), after noting that summer temperatures in North America are often higher than those in places where border-cuttings have been found useful, reported the results of a survey of regeneration in a stand of red spruce plus scattered white spruce that had been isolated by clearcutting on all sides, so furnishing an opportunity for observing regeneration on different exposures in this old-field stand at Dummer, New Hampshire. The regeneration included a surprisingly large number of balsam fir seedlings from the 5% stand component of that species. The maximum density of spruce regeneration, determined 4 rods (20 m) inside from the edge of the stand on a North 20°E exposure, was 600,000/ha, with almost 100,000 balsam fir seedlings.

A prepared seedbed remains receptive for a relatively short period, seldom as long as 5 years, sometimes as short as 3 years. Seedbed receptivity on moist, fertile sites decreases with particular rapidity, and especially on such sites, seedbed preparation should be scheduled to take advantage of good seed years. In poor seed years, site preparation can be carried out on mesic and drier sites with more chance of success, because of the generally longer receptivity of seedbeds there than those on moister sites. Although an indifferent seed year can suffice if seed distribution is good and environmental conditions favourable to seedling germination and survival, small amounts of seed are particularly vulnerable to depredation by small mammals. Considerable flexibility is possible in timing site preparation to coincide with cone crops. Treatment can be apllied either before any logging takes place, between partial cuts, or after logging. In cut and leave strips, seedbed preparation can be carried out as a single operation, pre-scarifying the leave strips, post-scarifying the cut strips.

Broadcast burning is not recommended as a method of preparing sites for natural regeneration, as it rarely exposes enough mineral soil to be sufficiently receptive, and the charred organic surfaces are a poor seedbed for spruce. A charred surface may get too hot for good germination and may delay germination until fall, with subsequent overwinter mortality of unhardened seedlings. Piling and burning of logging slash, however, can leave suitable exposures of mineral soil.

Season of Planting

Artificial Regeneration

With a view to reducing the time needed to produce planting stock, experiments were carried out with white spruce and three other coniferous species from Wisconsin seed in the longer, frost-free growing season in Florida, 125 vs. 265 days in central Wisconsin and northern Florida, respectively. As the species studied are adapted to long photoperiods, extended daylengths of 20 hours were applied in Florida. Other seedlings were grown under extended daylength in Wisconsin and with natural daylength in both areas. After two growing seasons, white spruce under long days in Florida were about the same as those in Wisconsin, but twice as tall as plants under natural Wisconsin photoperiods. Under natural days in Florida, with the short local photoperiod, white spruce was severely dwarfed and had a low rate of survival. Black spruce responded similarly. After two growing seasons, long day plants of all 4 species in Florida were well balanced, with good development of both roots and shoots, equaling or exceeding the minimum standards for 2+1 and 2+2 outplanting stock of Lake States species. Their survival when lifted in February and outplanted in Wisconsin equalled that of 2+2 Wisconsin-grown transplants. Artificial extension of the photoperiod in the northern Lake States greatly increased height increment of white and black spruces in the second growing season.

Optimum conditions for seedling growth have been determined for the production of containerized planting stock.Alternating day/night temperatures have been found more suitable than a constant temperature; at 400 lumens/m^2 light regime, a 28 °C/20 °C day/night temperatures have been recommended for white spruce. However, temperature optima are not necessarily the same at different ages and sizes. Tinus (1984) investigated the effects of combinations of day and night temperature on height, caliper, and dry weight of 4 seed sources of Engelmann spruce. The 4 seed sources appeared to have very similar temperature requirements, with night optima about the same of slightly lower than daylight optima.

Tree provenance is important in artificial regeneration. Good provenance takes into account suitable tree genetics and a good environmental fit for planted / seeded trees in a forest stand. The wrong genotype can lead to failed regeneration, or poor trees that are prone to pathogens and undesired outcomes.

Artificial regeneration has been a more common method involving planting because it is more dependable than natural regeneration. Planting can involve using seedlings (from a nursery), (un)rooted cuttings, or seeds.

Whichever method is chosen it can be assisted by tending techniques also known as intermediate stand treatments.

The fundamental genetic consideration in artificial regeneration is that seed and planting stock must be adapted to the planting environment. Most commonly, the method of managing seed and stock deployment is through a system of defined seed zones, within which seed and stock can be moved without risk of climatic maladaptation. Ontario adopted a seed zone system in the 1970s based on Hills' (1952) site regions and provincial resource district boundaries, but Ontario's seed zones are now based on homogeneous climatic regions developed with the Ontario Climate Model. The regulations stipulate that source-identified seedlots may be either a general collection, when only the seed zone of origin is known, or a stand collection from a specific latitude and longitude. The movement of general-collection seed and stock across seed zone boundaries is prohibited, but the use of stand-collection seed and stock in another seed zone is acceptable when the Ontario Climate Model shows that the planting site and place of seed origin are climatically similar. The 12 seed zones for white spruce in Quebec are based mainly on ecological regions, with a few modifications for administrative convenience.

Seed quality varies with source. Seed orchards produce seed of the highest quality, then, in order of decreasing seed quality produced, seed production areas and seed collection areas follow, with controlled general collections and uncontrolled general collections producing the least characterized seed.

Seeds

Dewinging, Extraction

When seed is first separated from cones it is mixed with foreign matter, often 2 to 5 times the volume of the seed. The more or less firmly attached membranous wings on the seed must be detached before it is cleaned of foreign matter. The testa must not incur damage during the dewinging process. Two methods have been used, dry and wet. Dry seed may be rubbed gently through a sieve that has a mesh through which only seed without wings can pass. Large quantities of seed can be processed in dewinging machines, which use cylinders of heavy wire mesh and rapidly revolving stiff brushes within to remove the wings. In the wet process, seed with wings attached are spread out 10 cm to 15 cm deep on a tight floor and slightly moistened throughout; light leather flails are used to free seed from the wings. Wang (1973) described a unique wet dewinging procedure using a cement mixer, used at the Petawawa tree seed processing facility. Wings of white and Norway spruce seed can be removed by dampening the seed slightly before it is run through a fanning mill for the last time. Any moistened seed must be dried before fermentation or moulding sets in.

Seed Viability

A fluorescein diacetate (FDA) biochemical viability test for several species of conifer seed, including white spruce, estimates the proportion of live seed (viability) in a seedlot, and hence the percentage germination of a seedlot. The accuracy of predicting percentage germination was within +/- 5 for most seedlots. White spruce seed can be tested for viability by an indirect method, such as the fluorescein diacetate (FDA) test or 'Ultra-sound'; or by the direct growth method of 'germination'. Samples of white spruce seed inspected by Toumey and Stevens (1928) varied in viability from 50% to 100%, but averaged 93%. Rafn (1915) reported 97% viability for white spruce seed.

Germinative Testing

The results of a germination test are commonly expressed as *germinative capacity* or a *germination percentage*, which is the percentage of seeds that germinate during a period of time, ending when germination is practically complete. During extraction and processing, white spruce seeds gradually lost moisture, and total germination increased. Mittal et al. (1987) reported that white spruce seed from Algonquin Park, Ontario, obtained the maximum rate (94% in 6 days) and 99% total germination in 21 days after 14-week pre-chilling. The pre-treatment of 1% sodium hypochlorite increased germinability.

Encouraged by Russian success in using ultrasonic waves to improve the germinative energy and percentage germination of seeds of agricultural crops, Timonin (1966) demonstrated benefits to white spruce germination after exposure of seeds to 1, 2, or 4 minutes of ultrasound generated by an M.S.E. ultrasonic disintegrator with a power consumption of 280 VA and power impact of 1.35 amperes.[Tables 3.18 and 3.19] However, no seeds germinated after 6 minutes of exposure to ultrasound.

Seed Dormancy

Seed dormancy is a complex phenomenon and is not always consistent within species. Cold stratification of white spruce seed to break dormancy has been specified as a requirement, but Heit (1961) and Hellum (1968) regarded stratification as unnecessary. Cone handling and storage conditions affect dormancy in that cold, humid storage (5 °C, 75% to 95% relative humidity) of the cones prior to extraction seemingly eliminated dormancy by overcoming the need to stratify.Periods of cold, damp weather during the period of cone storage might provide natural cold (stratification) treatment. Once dormancy was removed in cone storage, subsequent kiln-drying and seed storage did not reactivate dormancy.

Haddon and Winston (1982) found a reduction in viability of stratified seeds after 2 years of storage and suggested that stress might have been caused by stratification, e.g., by changes in seed biochemistry, reduced embryo vigor, seed aging or actual damage to the embryo. They further questioned the quality of the 2-year-old seed even though high germination occurred in the samples that were not stratified.

Cold Stratification

Cold stratification is the term applied to the storing of seeds in (and, strictly, in layers with) a moist medium, often peat or sand, with a view to maintaining viability and overcoming dormancy. Cold stratification is the term applied to storage at near-freezing temperatures, even if no medium is used. A common method of cold stratification, is to soak seed in tap water for up to 24 h, superficially dry it, then store moist for some weeks or even months at temperatures just above freezing. Although Hellum (1968) found that cold stratification of an Alberta seed source led to irregular germination, with decreasing germination with increasing length of the stratification period, Hocking's (1972) paired test with stratified and nonstratified Alberta seed from several sources revealed no trends in response to stratification. Hocking suggested that seed maturity, handling, and storage needed to be controlled before the need for stratification could be determined. Later,

Winston and Haddon (1981) found that the storage of white spruce cones for 4 weeks at 5 °C prior to extraction obviated the need for stratification.

Seed Ripeness

Seed maturity cannot be predicted accurately from cone flotation, cone moisture content, cone specific gravity; but the province of B.C. found embryo occupying 90% + of the corrosion cavity and megagametophyte being firm and whitish in colour are the best predictors for white spruce in B.C., and Quebec can forecast seed maturity some weeks in advance by monitoring seed development in relation to heat-sums and the phenological progression of the inflorescence of fireweed (*Epilobium angustifolium* L.), an associated plant species. Cone collection earlier than one week before seed maturity would reduce seed germination and viability during storage. Four stages of maturation were determined by monitoring carbohydrates, polyols, organic acids, respiration, and metabolic activity. White spruce seeds require a 6-week post-harvest ripening period in the cones to obtain maximum germinability, however, based on cumulative degree-days, seed from the same trees and stand showed 2-week cone storage was sufficient.

Forest Tree Nurseries

Forest tree Plantations

Plantation Establishment Criteria

Plantations may be considered successful when outplant performance satisfies certain criteria. The term "free growing" is applied in some jurisdictions. Ontario's "Free-to-Grow" (FTG) equivalent relates to a forest stand that meets a minimum stocking standard and height requirement, and is essentially free of competition from surrounding vegetation that might impede growth. The FTG concept was introduced with the advent of the Forest Management Agreement program in Ontario in 1980 and became applicable to all management units in 1986. Policy, procedures, and methodologies readily applicable by forest unit managers to assess the effectiveness of regeneration programs were still under development during the Class Environmental Assessment hearings.

In British Columbia, the Forest Practices Code (1995) governs performance criteria. To minimize the subjectivity of assessing deciduous competition as to whether or not a plantation is established, minimum specifications of number, health, height, and competition have been specified in British Columbia. However, minimum specifications are still subjectively set and may need to be fine-tuned in order to avoid unwarranted delay in according established status to a plantation. For example, a vigorous white spruce with a strong, multi-budded leading shoot and its crown fully exposed to light on 3 sides would not qualify as free-growing in the current British Columbia Code but would hardly warrant description as unestablished.

Competition

Competition arises when individual organisms are sufficiently close together to incur growth constraint through mutual modification of the local environment. Plants may compete for light, moisture and nutrients, but seldom for space *per se*. Vegetation management directs more of the site's resources into usable forest products, rather than just eliminating all competing plants. Ideally, site preparation ameliorates competition to levels that relieve the outplant of constraints severe enough to cause prolonged check.

The diversity of boreal and sub-boreal broadleaf-conifer mixed tree species stands, commonly referred to as the "mixedwoods", largely preclude the utility of generalizations and call for the development of management practices incorporating the greater inherent complexity of broadleaf-conifer mixtures, relative to single-species or mixed-species conifer forest. After harvesting or other disturbance, mixedwood stands commonly enter a prolonged period in which hardwoods overtop the coniferous component, subjecting them to intense competition in an understorey. It is well established that the regeneration and growth potential of understorey conifers in mixedwood stands is correlated to the density of competing hardwoods. To help apply "free-to-grow" regulations in British Columbia and Alberta, management guidelines based on distance-dependent relations within a limited radius of crop trees were developed, but Lieffers et al. (2002) found that free-growing stocking standards did not adequately characterize light competition between broadleaf and conifer components in boreal mixedwood stands, and further noted that adequate sampling using current approaches would be operationally prohibitive.

Many promising plantations have failed through lack of tending. Young crop trees are often ill-equipped to fight it out with competition resurgent following initial site preparation and planting.

Perhaps the most direct evaluation of the effect of competition on plantation establishment is provided by an effective herbicide treatment. The fact that herbicide treatment does not always produce positive results should not obscure the demonstrated potential of herbicides for significantly promoting plantation establishment. Factors that can vitiate the effectiveness of a herbicide treatment include: weather, especially temperature, prior to and during application; weather, especially wind, during application; weather, especially precipitation, in the 12 to 24 hours after application; vegetation characteristics, including species, size, shape, phenological stage, vigour, and distribution of weeds; crop characteristics, including species, phenology, and condition; the effects of other treatments, such as preliminary shearblading, burning or other prescribed or accidental site preparation; and the herbicide used, including dosage, formulation, carrier, spreader, and mode of application. There is a lot that can go wrong, but a herbicide treatment can be as good or better than any other method of site preparation.

Competition Indices

The study of competition dynamics requires both a measure of the competition level and a measure of crop response. Various competition indices have been developed, e.g., by Bella (1971) and Hegyi (1974) based on stem diameter, by Arney (1972), Ek and Monserud (1974), and Howard and Newton (1984) based on canopy development, and

Daniels (1976), Wagner (1982), and Weiner (1984) with proximity-based models. Studies generally considered tree response to competition in terms of absolute height or basal area, but Zedaker (1982) and Brand (1986) sought to quantify crop tree size and environmental influences by using relative growth measures.

Tending

Tending is the term applied to pre-harvest silvicultural treatment of forest crop trees at any stage after initial planting or seeding. The treatment can be of the crop itself (e.g., spacing, pruning, thinning, and improvement cutting) or of competing vegetation (e.g., weeding, cleaning).

Planting

How many trees per unit area (spacing) that should be planted is not an easily answered question. Establishment density targets or regeneration standards have commonly been based on traditional practice, with the implicit aim of getting the stand quickly to the free-to-grow stage. Money is wasted if more trees are planted than are needed to achieve desired stocking rates, and the chance to establish other plantations is proportionately diminished. Ingress (natural regeneration) on a site is difficult to predict and often becomes surprisingly evident only some years after planting has been carried out. Early stand development after harvesting or other disturbance undoubtedly varies greatly among sites, each of which has its own peculiar characteristics.

For all practical purposes, the total volume produced by a stand on a given site is constant and optimum for a wide range of density or stocking. It can be decreased, but not increased, by altering the amount of growing stock to levels outside this range. Initial density affects stand development in that close spacing leads to full site utilization more quickly than wider spacing. Economic operability can be advanced by wide spacing even if total production is less than in closely spaced stands.

Beyond the establishment stage, the relationship of average tree size and stand density is very important. Various density-management diagrams conceptualizing the density-driven stand dynamics have been developed. Smith and Brand's (1988) diagram has mean tree volume on the vertical axis and the number of trees/ha on the horizontal axis: a stand can either have many little trees or a few big ones. The self-thinning line shows the largest number of trees of a given size/ha that can be carried at any given time. However, Willcocks and Bell (1995) caution against using such diagrams unless specific knowledge of the stand trajectory is known.

In the Lake States, plantations have been made with the spacing between trees varying from 3 by 3 to 10 by 10 feet (0.9 m by 0.9 m to 3.0 m by 3.0 m). Kittredge recommended that no fewer than 600 established trees per acre (1483/ha) be present during the early life of a plantation. To insure this, at least 800 trees per acre (1077/ha) should be planted where 85% survival may be expected, and at least 1200/ac (2970/ha) if only half of them can be expected to live. This translates into recommended spacings of 5 by 5 to 8 by 8 feet (1.5 m by 1.5 m to 2.4 m by 2.4 m) for plantings of conifers, including white spruce in the Lake States.

Enrichment Planting

A strategy for enhancing natural forests' economic value is to increase their concentration of economically important, indigenous tree species by planting seeds or seedlings for future harvest, which can be accomplished with enrichment planting (EP). This means increasing the planting density (i.e., the numbers of plants per hectare) in an already growing forest stand."

Release Treatments

- Weeding: A process of getting rid of saplings' or seedlings' competition by mowing, application of herbicide, or other method of removal from the surroundings.
- Cleaning: Release of select saplings from competition by overtopping trees of a comparable age. The treatment favors trees of a desired species and stem quality.
- Liberation cutting: A treatment that releases tree seedling or saplings by removing older overtopping trees.

Spacing

Over-crowded regeneration tends to stagnate. The problem is aggravated in species that have little self-pruning ability, such as white spruce. Spacing is a thinning (of natural regeneration), in which all trees other than those selected for retention at fixed intervals are cut. The term juvenile spacing is used when most or all of the cut trees are unmerchantable. Spacing can be used to obtain any of a wide range of forest management objectives, but it is especially undertaken to reduce density and control stocking in young stands and prevent stagnation, and to shorten the rotation, i.e., to speed the production of trees of a given size. Volume growth of individual trees and the merchantable growth of stands are increased. The primary rationale for spacing is that thinning is the projected decline in maximum allowable cut. And since wood will be concentrated on fewer, larger, and more uniform stems, operating and milling costs will be minimized.

Methods for spacing may be: manual, using various tools, including power saws, brush saws, and clippers; mechanical, using choppersand mulchers; chemical; or combinations of several methods. One treatment has had notable success in spacing massively overstocked (<100 000 stems/ha) natural regeneration of spruce and fir in Maine. Fitted to helicopter, the Thru-Valve boom emits herbicide spray droplets 1000 μm to 2000 μm in diameter at very low pressure. Swaths 1.2 m wide and leave strips 2.4 m wide were obtained with "knife-edge" precision when the herbicide was applied by helicopter flying at a height of 21 m at a speed of 40–48 km/h. It seems likely that no other method could be as cost-effective.

Twenty years after spacing to 2.5 × 2.5 m, 30-year-old mixed stands of balsam fir and white spruce in the Green River watershed, New Brunswick, averaged 156.9 m^3/ha.

A spacing study of 3 conifers (white spruce, red pine and jack pine) was established at Moodie, Manitoba, on flat, sandy, nutritionally poor soils with a fresh moisture regime. Twenty years after planting, red pine had the largest average dbh, 15% greater than jack pine, while white spruce dbh was less than half that of the pines. Crown width showed a gradual increase with spacing for all 3 conifers. Results to date were suggesting optimum spacings between 1.8 m and 2.4 m for both pines; white spruce was not recommended for planting on such sites.

Comparable data are generated by espacement trials, in which trees are planted at a range of densities. Spacings of 1.25 m, 1.50 m, 1.75 m, 2.00 m, 2.50 m, and 3.00 m on 4 site classes were used in the 1922 trial at Petawawa, Ontario. In the first of 34 old field white spruce plantations used to investigate stand development in relation to spacing at Petawawa, Ontario, regular rows were planted at average spacings of from 4 × 4 to 7 × 7 feet (1.22 m × 1.22 m to 2.13 m × 2.13 m). Spacings up to 10 × 10 feet (3.05 m × 3.03 m) were subsequently included in the study. Yield tables based on 50 years of data showed:

a) Except for merchantable volumes at age 20 and site classes 50 and 60, closer spacings gave greater standing volumes at all ages than did wider spacings, the relative difference decreasing with age.
b) Merchantable volume as a proportion of total volume increases with age, and is greater at wider than at closer spacings.
c) Current annual volume increment culminates sooner at closer than at wider spacings.

A smaller espacement trial, begun in 1951 near Thunder Bay, Ontario, included white spruce at spacings of 1.8 m, 2.7 m, and 3.6 m. At the closest spacing, mortality had begun at 37 years, but not at the wider spacings.

The oldest interior spruce espacement trial in British Columbia was established in 1959 near Houston in the Prince Rupert Forest Region. Spacings of 1.2 m, 2.7 m, 3.7 m, and 4.9 m were used, and trees were measured 6, 12, 16, 26, and 30 years after planting. At wide espacements, trees developed larger diameters, crowns, and branches, but (at 30 years) basal area and total volume/ha were greatest in the closest espacement (Table 6.38). In more recent trials in the Prince George Region of British Columbia (Table 6.39) and in Manitoba, planting density of white spruce had no effect on growth after up to 16 growing seasons, even at spacings as low as 1.2 m. The slowness of juvenile growth and of crown closure delay the response to intra-competition. Initially, close spacing might even provide a positive nurse effect to offset any negative response to competition.

Thinning

Thinning is an operation that artificially reduces the number of trees growing in a stand with the aim of hastening the development of the remainder. The goal of thinning is to control the amount and distribution of available growing space. By altering stand density, foresters can influence the growth, quality, and health of residual trees. It also provides an opportunity to capture mortality and cull the commercially less desirable, usually smaller and malformed, trees. Unlike regeneration treatments, thinnings are not intended to establish a new tree crop or create permanent canopy openings.

Thinning greatly influences the ecology and micro-meteorology of the stand, lowering the inter-tree competition for water. The removal of any tree from a stand has repercussions on the remaining trees both above-ground and below. Silvicultural thinning is a powerful tool that can be used to influence stand development, stand stability, and the characteristics of the harvestable products.

When considering intensive conifer plantations designed for maximum production, it is essential to remember that tending and thinning regimes and wind and snow damage are intimately related.

Previous studies have demonstrated that repeated thinnings over the course of a forest rotation increase carbon stores relative to stands that are clear-cut on short rotations and that the carbon benefits differ according to thinning method (e.g., thinning from above versus below).

Precommercial Thinning

When natural regeneration or artificial seeding has resulted in dense, overstocked young stands, natural thinning will in most cases eventually reduce stocking to more silviculturally desirable levels. But by the time some trees reach merchantable size, others will be overmature and defective, and others will still be unmerchantable. The yield of merchantable wood can be greatly increased and the rotation shortened by precommercial thinning. Mechanical and chemical methods have been applied, but their costliness has militated against their ready adoption.

Pruning

Pruning, as a silvicultural practice, refers to the removal of the lower branches of the young trees (also giving the shape to the tree) so clear knot-free wood can subsequently grow over the branch stubs. Clear knot-free lumber has a higher value. Pruning has been extensively carried out in the Radiata pine plantations of New Zealand and Chile, however the development of Finger joint technology in the production of lumber and mouldings has led to many forestry companies reconsidering their pruning practices. "Brashing" is an alternative name for the same process. Pruning can be done to all trees, or more cost effectively to a limited number of trees. There are two types of pruning: natural or self-pruning and artificial pruning. Most cases of self-pruning happen when branches do not receive enough sunlight and die. Wind can also take part in natural pruning which can break branches. Artificial pruning is where people are paid to come and cut the branches. Or it can be natural, where trees are planted close enough that the effect is to cause self-pruning of low branches as energy is put into growing up for light reasons and not branchiness.

Stand Conversion

The term *stand conversion* refers to a change from one silvicultural system to another and includes *species conversion*, i.e., a change from one species (or set of species) to another. Such change can be effected intentionally by various silvicultural means, or incidentally by default e.g., when high-grading has removed the coniferous content from a mixedwood stand, which then becomes exclusively self-perpetuating aspen. In general, such sites as these are the most likely to be considered for conversion.

Growth and Yield

In discussing yields that might be expected from the Canadian spruce forests, Haddock (1961) noted that Wright's (1959) quotation of spruce yields in the British Isles of 220 cubic feet per acre (15.4 m^3/ha) per year and in Germany of 175 cubic feet per acre (12.25 m^3/ha) per year was misleading, at least if it was meant to imply that such yields might be approached in the Boreal Forest Region of Canada. Haddock thought that Wright's suggestion of 20 to 40 (average 30) cubic feet per acre (1.4 m^3/ha to 2.8 m^3/ha (average 2.1 m^3/ha) per year was more reasonable, but still somewhat optimistic.

The principal way forest resource managers influence growth and yield is to manipulate the mixture of species and number (density) and distribution (stocking) of individuals that form the canopy of the stand. Species composition of much of the boreal forest in North America already differs greatly from its pre-exploitation state. There is less spruce and more hardwoods in the second-growth forest than in the original forest; Hearnden et al. (1996) calculated that the spruce cover type had declined from 18% to only 4% of the total forested area in Ontario. Mixedwood occupies a greater proportion of Ontario's second-growth forest (41%) than in the original (36%), but its component of white spruce is certainly much diminished.

Growth performance is certainly influenced by site conditions and thus by the kind and degree of site preparation in relation to the nature of the site. It is important to avoid the assumption that site preparation of a particular designation will have a particular silvicultural outcome. Scarification, for instance, not only covers a wide range of operations that scarify, but also any given way of scarifying can have significantly different results depending on site conditions at the time of treatment. In point of fact, the term is commonly misapplied. *Scarification* is defined as "Loosening the top soil of open areas, or breaking up the forest floor, in preparation for regenerating by direct seeding or natural seedfall", but the term is often misapplied to practices that include scalping, screefing, and blading, which pare off low and surface vegetation, together with most off its roots to expose a weed-free surface, generally in preparation for sowing or planting thereon.

Thus, it is not surprising that literature can be used to support the view that the growth of seedlings on scarified sites is much superior to that of growth on similar sites that have not been scarified, while other evidence supports the contrary view that scarification can reduce growth. Detrimental results can be expected from scarification that impoverishes the rooting zone or exacerbates edaphic or climatic constraints.

Burning site preparation has enhanced spruce seedling growth, but it must be supposed that burning could be detrimental if the nutrient capital is significantly depleted.

An obvious factor greatly influencing regeneration is competition from other vegetation. In a pure stand of Norway spruce, for instance, Roussel (1948) found the following relationships:

Percent cover (%)	Vegetation Description
Below 1	No vegetation
1-3	Moss carpet with a few fir seedlings

4-10	Herbaceous plants appear
10-25	Bramble, herbs, fairly vigorous spruce seedlings
>25	Herbs, brambles very dense, vigorous, no moss

A factor of some importance in solar radiation–reproduction relationships is excess heating of the soil surface by radiation.This is especially important for seedlings, such as spruce, whose first leaves do not shade the base of the stem at the soil surface. Surface temperatures in sandy soils on occasion reach lethal temperatures of 50 °C to 60 °C.

Common Methods of Harvesting

Silvicultural regeneration methods combine both the harvest of the timber on the stand and re-establishment of the forest. The proper practice of sustainable forestry should mitigate the potential negative impacts, but all harvest methods will have some impacts on the land and residual stand. The practice of sustainable forestry limits the impacts such that the values of the forest are maintained in perpetuity. Silvicultural prescriptions are specific solutions to a specific set of circumstances and management objectives. Following are some common methods:

Clearcut Harvesting

Conventional clearcut harvesting is relatively simple: all trees on a cutblock are felled and bunched with bunches aligned to the skidding direction, and a skidder then drags the bunches to the closest log deck. Feller-buncher operators concentrate on the width of the felled swath, the number of trees in a bunch, and the alignment of the bunch. Providing a perimeter boundary is felled during daylight, night-shift operations can continue without the danger of trespassing beyond the block. Productivity of equipment is maximized because units can work independently of one another.

Clearcutting

An even-aged regeneration method that can employ either natural or artificial regeneration. It involves the complete removal of the forest stand at one time. Clearcutting can be biologically appropriate with species that typically regenerate from stand replacing fires or other major disturbances, such as Lodgepole Pine (*Pinus contorta*). Alternatively, clearcutting can change the dominating species on a stand with the introduction of non-native and invasive species as was shown at the Blodgett Experimental Forest near Georgetown California. Additionally, clearcutting can prolong slash decomposition, expose soil to erosion, impact visual appeal of a landscape and remove essential wildlife habitat. It is particularly useful in regeneration of tree species such as Douglas-fir (*Pseudotsuga menziesii*) which is shade intolerant. In addition, the general public's distaste for even-aged silviculture, particularly clearcutting, is likely to result in a greater role for uneven-aged management on public lands as well. Across Europe, and in parts of North America, even-aged, production-orientated and intensively managed plantations are beginning to be regarded in the same way as old industrial complexes: something to abolish or convert to something else.

Clearcutting will impact many site factors important in their effect on regeneration, including air and soil temperatures. Kubin and Kemppainen (1991), for instance, measured temperatures in northern Finland from 1974 through 1985 in 3 clear-felled areas and in 3 neighouring forest stands dominated by Norway spruce. Clear felling had no significant influence on air temperature at 2 m above the ground surface, but the daily air temperature maxima at 10 cm were greater in the clear-felled area than in the uncut forest, while the daily minima at 10 cm were lower. Night frosts were more common in the clear-felled area. Daily soil temperatures at 5 cm depth were 2 °C to 3 °C greater in the clear-felled area than in the uncut forest, and temperatures at depths of 50 cm and 100 cm were 3 °C to 5 °C greater. The differences between the clear-felled and uncut areas did not diminish during the 12 years following cutting.

Coppicing

A regeneration method which depends on the sprouting of cut trees. Most hardwoods, the coast redwood, and certain pines naturally sprout from stumps and can be managed through coppicing. Coppicing is generally used to produce fuelwood, pulpwood, and other products dependent on small trees. A close relative of coppicing is pollarding. Three systems of coppice woodland management are generally recognized: simple coppice, coppice with standards, and the coppice selection system.

- In Compound coppicing or **coppicing with standards**, some of the highest quality trees are retained for multiple rotations in order to obtain larger trees for different purposes.

Direct Seeding

Prochnau (1963), 4 years after sowing, found that 14% of viable white spruce seed sown on mineral soil had produced surviving seedlings, at a seed:seedling ratio of 7.1:1. With Engelmann spruce, Smith and Clark (1960) obtained average 7th year seed:seedling ratios of 21:1 on scarified seedbeds on dry sites, 38:1 on moist sites, and 111:1 on litter seedbeds.

Group Selection

The group selection method is an uneven-aged regeneration method that can be used when mid-tolerant species regeneration is desired. The group selection method can still result in residual stand damage in dense stands, however directional falling can minimize the damage. Additionally, foresters can select across the range of diameter classes in the stand and maintain a mosaic of age and diameter classes.

Méthode du contrôle

Classical European silviculture achieved impressive results with systems such as Henri Biolley's *méthode du contrôle* in Switzerland, in which the number and size of trees harvested were determined by reference to data collected from every tree in every stand measured every 7 years.

While not designed to be applied to boreal mixedwoods, the *méthode du contrôle* is described briefly here to illustrate the degree of sophistication applied by some European foresters to the management of their forests. Development of management techniques that allowed for stand development to be monitored and guided into sustainable paths were in part a response to past experience, particularly in Central European countries, of the negative effects of pure, uniform stands with species often unsuited to the site, which greatly increased the risk of soil degradation and biotic diseases. Increased mortality and decreased increment generated widespread concern, especially after reinforcement by other environmental stresses.

More or less uneven-aged, mixed forests of preponderantly native species, on the other hand, treated along natural lines, have proved to be healthier and more resistant to all kinds of external dangers; and in the long run such stands are more productive and easier to protect.

However, irregular stands of this type are definitely more difficult to manage—new methods and techniques had to be sought particularly for the establishment of inventories, as well as increment control and yield regulation. In Germany, for instance, since the beginning of the nineteenth century under the influence of G.L. Hartig (1764–1837), yield regulation has been effected almost exclusively by allotment or formula methods based on the conception of the uniform normal forest with a regular succession of cutting areas.

In France, on the other hand, efforts were made to apply another kind of forest management, one that aimed to bring all parts of the forest to a state of highest productive capacity in perpetuity. In 1878, the French forester A. Gurnaud (1825–1898) published a description of a *méthode du contrôle* for determining increment and yield. The method was based on the fact that through careful, selective harvesting, the productivity of the residual stand can be improved, because timber is removed as a cultural operation. In this method, the increment of stands is accurately determined periodically with the object of gradually converting the forest, through selective management and continuous experimentation, to a condition of equilibrium at maximum productive capacity.

Henri Biolley (1858–1939) was the first to apply Gurnaud's inspired ideas to practical forestry. From 1890 on, he managed the forests of his Swiss district according to these principles, devoting himself for almost 50 years to the study of increment and a treatment of stands directed towards the highest production, and proving the practicability of the check method. In 1920, he published this study giving a theoretical basis of management of forests under the check method, describing the procedures to be applied in practice (which he partly developed and simplified), and evaluating the results.

Biolley's pioneering work formed the basis upon which most Swiss forest management practices were later developed, and his ideas have been generally accepted. Today, with the trend of intensifying forest management and productivity in most countries, the ideas and application of careful, continuous treatment of stands with the aid of the volume check method are meeting with ever-growing interest. In Britain and Ireland, for example, there is increased application of Continuous Cover Forestry principles to create permanently irregular structures in many woodlands.

Patch cut

Row and Broadcast Seeding

Spot and row seeders use less seed that does broadcast ground or aerial seeding but may induce clumping. Row and spot seeding confer greater ability to control seed placement than does broadcast seeding. Also, only a small percentage of the total area needs to be treated.

In the aspen type of the Great Lakes region, direct sowing of the seed of conifers has usually failed. However, Gardner (1980) after trials in Yukon, which included broadcast seeding of white spruce seed at 2.24 kg/ha that secured 66.5% stocking in the Scarified Spring Broadcast treatment 3 years after seeding, concluded that the technique held "considerable promise".

Seed-tree

An even-aged regeneration method that retains widely spaced residual trees in order to provide uniform seed dispersal across a harvested area. In the seed-tree method, 2-12 seed trees per acre (5-30/ha) are left standing in order to regenerate the forest. They will be retained until regeneration has become established at which point they may be removed. It may not always be economically viable or biologically desirable to re-enter the stand to remove the remaining seed trees. Seed-tree cuts can also be viewed as a clearcut with natural regeneration and can also have all of the problems associated with clearcutting. This method is most suited for light-seeded species and those not prone to windthrow.

Selection Systems

Selection systems are appropriate where uneven stand structure is desired, particularly where the need to retain continuous cover forest for aesthetic or environmental reasons outweighs other management considerations. Selection logging has been suggested as being of greater utility than shelterwood systems in regenerating old-growth Engelmann Spruce Sub-alpine Fir (ESSF) stands in southern British Columbia. In most areas, selection logging favours regeneration of fir more than the more light-demanding spruce. In some areas, selection logging can be expected to favour spruce over less tolerant hardwood species (Zasada 1972) or lodgepole pine.

Shelter Spot Seeding

The use of shelters to improve germination and survival in spot seedings seeks to capture the benefits of greenhouse culture, albeit miniature. The Hakmet seed shelter, for instance, is a semi-transparent plastic cone 8 cm high, with openings of 7 cm diameter in the 7.5 cm diameter base and 17 mm diameter in the 24 mm diameter top. This miniature greenhouse increases air humidity, reduces soil desiccation, and raises air and soil temperatures to levels more favourable to germination and seedling growth than those offered by unprotected conditions. The shelter is designed to break down after a few years of exposure to ultraviolet radiation.

Seed shelters and spring sowing significantly improved stocking compared with bare spot seeding, but sheltering did not significantly improve growth. Stocking of bare seedspots was extremely low, possibly due to smothering of seedlings by abundant broadleaf and herbaceous litter, particularly that from aspen and red raspberry, and exacerbated by strong competition from graminoids and raspberry.

Cone shelters (Cerkon™) usually produced greater survival than unsheltered seeding on scarified seedspots in trials of direct seeding techniques in interior Alaska, and funnel shelters (Cerbel™) usually produced greater survival than unsheltered seeding on non-scarified seedspots. Both shelter types are manufactured by AB Cerbo in Trollhättan, Sweden. Both are made of light-degradable, white, opaque plastic, and are 8 cm high when installed.

White spruce seed was sown in Alaska on a burned site in summer 1984, and protected by white plastic cones on small spots scarified by hand, or by white funnels placed directly into the residual ash and organic material. A group of 6 ravens (*Corvus corax*) was observed in the area about 1 week after sowing was completed in mid-June. Damage averaged 68% with cones and 50% with funnels on an upland area, and 26% with funnels on a floodplain area. Damage by ravens was only 0.13% on unburned but otherwise similar areas.

In seeding trials in Manitoba between 1960 and 1966 aimed at converting aspen stands to spruce–aspen mixedwoods, 1961 scarification in the Duck Mountain Provincial Forest remained receptive to natural seeding for many years.

Shelterwood

In general terms, the shelterwood system is a series of partial cuts that removes the trees of an existing stand over several years and eventually culminates in a final cut that creates a new even-aged stand. It is an even-aged regeneration method that removes trees in a series of three harvests: 1) Preparatory cut; 2) Establishment cut; and 3) Removal cut. The success of practising a shelterwood system is closely related to: 1. the length of the regeneration period, i.e. the time from the shelterwood cutting to the date when a new generation of trees has been established; 2.the quality of the new tree stand with respect to stand density and growth; and 3.the value increment of the shelter trees. Information on the establishment, survival and growth of seedlings influenced by the cover of shelter trees, as well as on the growth of these trees, is needed as a basis for modelling the economic return of practising a shelterwood system. The method's objective is to establish new forest reproduction under the shelter of the retained trees. Unlike the seed-tree method, residual trees alter understory environmental conditions (i.e. sunlight, temperature, and moisture) that influence tree seedling growth. This method can also find a middle ground with the light ambiance by having less light accessible to competitors while still being able to provide enough light for tree regeneration. Hence, shelterwood methods are most often chosen for site types characterized by extreme conditions, in order to create a new tree generation within a reasonable time period. These conditions are valid foremost on level ground sites which are either dry and poor or moist and fertile.

Shelterwood Systems

Shelterwood systems involve 2, 3, or exceptionally more partial cuttings. A final cut is made once adequate natural regeneration has been obtained. The shelterwood system is most commonly applied as a 2-cut uniform shelterwood, first an initial regeneration (seed) cut, the second a final harvest cut. In stands less than 100 years old, a light preparatory cut can be useful. A series of intermediate cuts at intervals of 10–20 years has been recommended for intensively managed stands.

From operational or economic standpoints, however, there are disadvantages to the shelterwood system: harvesting costs are higher; trees left for deferred cutting may be damaged during the regeneration cut or related extraction operations; the increased risk of blowdown threatens the seed source; damage from bark beetles is likely to increase; regeneration may be damaged during the final cut and related extraction operations; the difficulty of any site preparation would be increased; and incidental damage to regeneration might be caused by any site preparation operations.

Single-Tree Selection

The single-tree selection method is an uneven-aged regeneration method most suitable when shade tolerant species regeneration is desired. It is typical for older and diseased trees to be removed, thus thinning the stand and allowing for younger, healthy trees to grow. Single-tree selection can be very difficult to implement in dense or sensitive stands and residual stand damage can occur. This method is also disturbs the canopy layer the least out of all other methods.

Spot Seeding

Spot seeding was found to be the most economical and reliable of the direct seeding methods for converting aspen and paper birch to spruce and pine. In the Chippewa National Forest (Lake States), seed-spot sowing of 10 seeds each of white spruce and white pine under 40-year aspen after different degrees of cutting on gave second-season results clearly indicating the need to remove or disturb the forest floor to obtain germination of seeded white spruce and white pine.

Spot seeding of coniferous seed, including white spruce, has had occasional success, but several constraining factors commonly limit germination success: the drying out of the forest floor before the roots of germinants reach underlying moisture reserves; and, particularly under hardwoods, the smothering of small seedlings by snow-pressed leaf litter and lesser vegetation. Kittredge and Gervorkiantz (1929) determined that removal of the aspen forest floor increased germination percentage after the second season in seed spots of both white pine and white spruce, in 4 plots, from 2.5% to 5%, from 8% to 22%, from 1% to 9.5%, and from 0% to 15%.

Spot seeding requires less seed than broadcast seeding and tends to achieve more uniform spacing, albeit sometimes with clumping. The devices used in Ontario for manual spot seeding are the “oil can” seeder, seeding sticks, and shakers. The oil can is a container fitted with a long spout through which a predetermined number of seeds are released with each flick of the seeder.

Strip Cutting

Harvesting cutblocks where only a portion of the trees are to be removed is very different from clearcutting. First, trails must be located to provide access for the felling and skidding/forwarding equipment. These trails must be carefully located to ensure that the trees remaining meet the desired quality criteria and stocking density. Second, the equipment must not damage the residual stand. The further desiderata are outlined by Sauder (1995).

The dearth of seed and a deficiency of receptive seedbeds were recognized as major reasons for the lack of success of clearcut harvesting. One remedy attempted in British Columbia and Alberta has been alternate strip cutting. The greater seed source from uncut trees between the cut strips, and the disturbance to the forest floor within the cut strips could be expected to increase the amount of natural regeneration. Trees were cut to a diameter limit in the cut strips, but large trees in the leave strips often proved too much of a temptation and were cut too, thus removing those trees that would otherwise have been the major source of seed.

An unfortunate consequence of strip thinning was the build-up of spruce beetle populations. Shaded slash from the initial cut, together with an increase in the number of windthrown trees in the leave strips, provided conditions ideally suited to the beetle.

Underplanting

DeLong et al. (1991) suggested underplanting 30- to 40-year-old aspen stands, on the basis of the success of natural spruce in regenerating under stands of such stands: "By planting, spacing can be controlled enabling easier protection of the spruce during stand entry for harvesting of the aspen overstorey".

Variable Retention

A harvesting and regeneration method which is a relatively new silvicultural system that retains forest structural elements (stumps, logs, snags, trees, understory spieces and undisturbed layers of forest floor) for at least one rotation in order to preserve environmental values associated with structurally complex forests.

"Uneven-aged and even-aged methods differ in the scale and intensity of disturbance. Uneven-aged methods maintain a mix of tree sizes or ages within a habitat patch by periodically harvesting individual or small groups of trees, Even-aged methods harvest most or all of the overstory and create a fairly uniform habitat patch dominated by trees of the same age". Even-aged management systems have been the prime methods to use when studying the effects on birds.

Mortality

A survey in 1955–56 to determine survival, development, and the reasons for success or failure of conifer pulpwood plantations (mainly of white spruce) in Ontario and Quebec up to 32 years old found that the bulk of the mortality occurred within the first 4 years of planting, unfavourable site and climate being the main causes of failure.

Advance Growth

Naturally regenerated trees in an understorey prior to harvesting constitute a classic case of good news and bad news. Understorey white spruce is of particular importance in mixedwoods dominated by aspen, as in the B15, B18a, and B19a Sections of Manitoba, and elsewhere. Until the latter part of the last century, white spruce understorey was mostly viewed as money in the bank on a long-term, low interest deposit, with final yield to be realized after slow natural succession, but the resource became increasingly threatened with the intensification of harvesting of aspen. White spruce plantations on mixedwood sites proved expensive, risky, and generally unsuccessful. This prompted efforts to see what might be done about growing aspen and white spruce on the same landbase by protecting existing white spruce advance growth, leaving a range of viable crop trees during the first cut, then harvesting both hardwoods and spruce in the final cut. Information about the understorey component is critical to spruce management planning. The ability of then current harvesting technology and crews employed to provide adequate protection for white spruce understories was questioned by Brace and Bella. Specialized equipment and training, perhaps with financial incentives, may be needed to develop procedures that would confer the degree of protection needed for the system to be feasible. Effective understorey management planning requires more than improved mixedwood inventory.

Avoidance of damage to the understorey will always be a desideratum. Sauder's (1990) paper on mixedwood harvesting describes studies designed to evaluate methods of reducing non-trivial damage to understorey residuals that would compromise their chance of becoming a future crop tree. Sauder concluded that: (1) operational measures that protected residual stems may not unduly increase costs, (2) all felling, conifers and hardwoods, needs to be done in one operation to minimize the entry of the feller-buncher into the residual stand, (3) several operational procedures can reduce understorey damage, some of them without incurring extra costs, and (4) successful harvesting of treatment blocks depends primarily on the intelligent location of skid trails and landings. In summary, the key to protecting the white spruce understorey without sacrificing logging efficiency is a combination of good planning, good supervision, the use of appropriate equipment, and having conscientious, well-trained operators.Even the best plan will not reduce understorey damage unless its implementation is supervised.

New stands need to be established to provide for future supply of commercial white spruce from 150 000 ha of boreal mixedwoods in 4 of Rowe's (1972) regional Forest Sections straddling Alberta, Saskatchewan, and Manitoba, roughly from Peace River AB to Brandon MB. In the 1980s, with harvesting using conventional equipment and procedures, a dramatic increase in the demand for aspen posed a serious problem for the associated spruce understorey. Formerly, white spruce in the understories had developed to commercial size through natural succession under the protection of the hardwoods. Brace articulated a widespread concern: "The need for protection of spruce as a component of boreal mixedwoods goes beyond concern for the future commercial softwood timber supply. Concerns also include fisheries and wildlife habitat, aesthetics and recreation, a general dissatisfaction with cleacutting in mixedwoods and a strong interest in mixedwood perpetuation, as expressed recently in 41 public meetings on forestry development in northern Alberta...".

On the basis of tests of 3 logging systems in Alberta, Brace (1990) affirmed that significant amounts of understorey can be retained using any of those systems provided that sufficient effort is directed towards protection. Potential benefits would include increased short-term softwood timber supply, improved wildlife habitat and cutblock aesthetics, as well as reduced public criticism of previous logging practices. Stewart et al. (2001) developed statistical models to predict the natural establishment and height growth of understorey white spruce in the boreal mixedwood forest in Alberta using data from 148 permanent sample plots and supplementary information about height growth of white spruce regeneration and the amount and type of available substrate. A discriminant model correctly classified 73% of the sites as to presence or absence of a white spruce understorey, based on the amount of spruce basal area, rotten wood, ecological nutrient regime, soil clay fraction, and elevation, although it explained only 30% of the variation in the data. On sites with a white spruce understorey, a regression model related the abundance of regeneration to rotten wood cover, spruce basal area, pine basal area, soil clay fraction, and grass cover (R^2 = 0.36). About half the seedlings surveyed grew on rotten wood, and only 3% on mineral soil, and seedlings were 10 times more likely to have established on these substrates than on litter. Exposed mineral soil covered only 0.3% of the observed transect area.

Advance Growth Management

Advance growth management, i.e., the use of suppressed understorey trees, can reduce reforestation costs, shorten rotations, avoid denuding the site of trees, and also reduce adverse impacts on aesthetic, wildlife, and watershed values. To be of value, advance growth must have acceptable species composition and distribution, have potential for growth following release, and not be vulnerable to excessive damage from logging.

The age of advance growth is difficult to estimate from its size, as white that appears to be 2- to 3-year-old may well be more than 20 years old. However, age does not seem to determine the ability of advance growth of spruce to respond to release, and trees older than 100 years have shown rapid rates of growth after release. Nor is there a clear relationship between the size of advance growth and its growth rate when released.

Where advance growth consists of both spruce and fir, the latter is apt to respond to release more quickly than the former, whereas spruce does respond. If the ratio of fir to spruce is large, however, the greater responsiveness to release of fir may subject the spruce to competition severe enough to negate much of the effect of release treatment. Even temporary relief from shrub competition has increased height growth rates of white spruce in northwestern New Brunswick, enabling the spruce to overtop the shrubs.

Site Preparation

Site preparation is any of various treatments applied to a site in order to ready it for seeding or planting. The purpose is to facilitate the regeneration of that site by the chosen method. Site preparation may be designed to achieve, singly or in any combination: improved access, by reducing or rearranging slash, and amelioration of adverse forest floor, soil, vegetation, or other biotic factors. Site preparation is undertaken to ameliorate one or more constraints that would otherwise be likely to thwart the objectives of management. A valuable bibliography on the effects of soil temperature and site preparation on subalpine and boreal tree species has been prepared by McKinnon et al. (2002).

Site preparation is the work that is done before a forest area is regenerated. Some types of site preparation are burning.

Burning

Broadcast burning is commonly used to prepare clearcut sites for planting, e.g., in central British Columbia, and in the temperate region of North America generally.

Prescribed burning is carried out primarily for slash hazard reduction and to improve site conditions for regeneration; all or some of the following benefits may accrue:

- Reduction of logging slash, plant competition, and humus prior to direct seeding, planting, scarifying or in anticipation of natural seeding in partially cut stands or in connection with seed-tree systems.
- Reduction or elimination of unwanted forest cover prior to planting or seeding, or prior to preliminary scarification thereto.
- Reduction of humus on cold, moist sites to favour regeneration.
- Reduction or elimination of slash, grass, or brush fuels from strategic areas around forested land to reduce the chances of damage by wildfire.

Prescribed burning for preparing sites for direct seeding was tried on a few occasions in Ontario, but none of the burns was hot enough to produce a seedbed that was adequate without supplementary mechanical site preparation.

Changes in soil chemical properties associated with burning include significantly increased pH, which Macadam (1987) in the Sub-boreal Spruce Zone of central British Columbia found persisting more than a year after the burn. Average fuel consumption was 20 to 24 t/ha and the forest floor depth was reduced by 28% to 36%. The increases correlated well with the amounts of slash (both total and $\geq$7 cm diameter) consumed. The change in pH depends on the severity of the burn and the amount consumed; the increase can be as much as 2 units, a 100-fold change. Deficiencies of copper and iron in the foliage of white spruce on burned clearcuts in central British Columbia might be attributable to elevated pH levels.

Even a broadcast slash fire in a clearcut does not give a uniform burn over the whole area. Tarrant (1954), for instance, found only 4% of a 140-ha slash burn had burned severely, 47% had burned lightly, and 49% was unburned. Burning after windrowing obviously accentuates the subsequent heterogeneity.

Marked increases in exchangeable calcium also correlated with the amount of slash at least 7 cm in diameter consumed.Phosphorus availability also increased, both in the forest floor and in the 0 cm to 15 cm mineral soil layer, and the increase was still evident, albeit somewhat diminished, 21 months after burning. However, in another study in the same Sub-boreal Spruce Zone found that although it increased immediately after the burn, phosphorus availability had dropped to below pre-burn levels within 9 months.

Nitrogen will be lost from the site by burning, though concentrations in remaining forest floor were found by Macadam (1987) to have increased in 2 of 6 plots, the others showing decreases. Nutrient losses may be outweighed, at least in the short term, by improved soil microclimate through the reduced thickness of forest floor where low soil temperatures are a limiting factor.

The *Picea/Abies* forests of the Alberta foothills are often characterized by deep accumulations of organic matter on the soil surface and cold soil temperatures, both of which make reforestation difficult and result in a general deterioration in site productivity; Endean and Johnstone (1974) describe experiments to test prescribed burning as a means of seedbed preparation and site amelioration on representative clear-felled *Picea/Abies* areas. Results showed that, in general, prescribed burning did not reduce organic layers satisfactorily, nor did it increase soil temperature, on the sites tested. Increases in seedling establishment, survival, and growth on the burned sites were probably the result of slight reductions in the depth of the organic layer, minor increases in soil temperature, and marked improvements in the efficiency of the planting crews. Results also suggested that the process of site deterioration has not been reversed by the burning treatments applied.

Ameliorative Intervention

Slash weight (the oven-dry weight of the entire crown and that portion of the stem < 4 inches in diameter) and size distribution are major factors influencing the forest fire hazard on harvested sites. Forest managers interested in the application of prescribed burning for hazard reduction and silviculture, were shown a method for quantifying the slash load by Kiil (1968).In west-central Alberta, he felled, measured, and weighed 60 white spruce, graphed (a) slash weight per merchantable unit volume against diameter at breast height (dbh), and (b) weight of fine slash (<1.27 cm) also against dbh, and produced a table of slash weight and size distribution on one acre of a hypothetical stand of white spruce. When the diameter distribution of a stand is unknown, an estimate of slash weight and size distribution can be obtained from average stand diameter, number of trees per unit area, and merchantable cubic foot volume. The sample trees in Kiil's study had full symmetrical crowns. Densely growing trees with short and often irregular crowns would probably be overestimated; open-grown trees with long crowns would probably be underestimated.

The need to provide shade for young outplants of Engelmann spruce in the high Rocky Mountains is emphasized by the U.S. Forest Service. Acceptable planting spots are defined as microsites on the north and east sides of down logs, stumps, or slash, and lying in the shadow cast by such material. Where the objectives of management specify more uniform spacing, or higher densities, than obtainable from an existing distribution of shade-providing material, redistribution or importing of such material has been undertaken.

Access

Site preparation on some sites might be done simply to facilitate access by planters, or to improve access and increase the number or distribution of microsites suitable for planting or seeding.

Wang et al. (2000) determined field performance of white and black spruces 8 and 9 years after outplanting on boreal mixedwood sites following site preparation (Donaren disc trenching versus no trenching) in 2 plantation types (open versus sheltered) in southeastern Manitoba. Donaren trenching slightly reduced the mortality of black spruce but significantly increased the mortality of white spruce. Significant difference in height was found between open and sheltered plantations for black spruce but not for white spruce, and root collar diameter in sheltered plantations was significantly larger than in open plantations for black spruce but not for white spruce. Black spruce open plantation had significantly smaller volume (97 cm^3) compared with black spruce sheltered (210 cm^3), as well as white spruce open (175 cm^3) and sheltered (229 cm^3) plantations. White spruce open plantations also had smaller volume than white spruce sheltered plantations. For transplant stock, strip plantations had a significantly higher volume (329 cm^3) than open plantations (204 cm^3). Wang et al. (2000) recommended that sheltered plantation site preparation should be used.

Mechanical

Up to 1970, no "sophisticated" site preparation equipment had become operational in Ontario, but the need for more efficacious and versatile equipment was increasingly recognized. By this time, improvements were being made to equipment originally developed by field staff, and field testing of equipment from other sources was increasing.

According to J. Hall (1970), in Ontario at least, the most widely used site preparation technique was post-harvest mechanical scarification by equipment front-mounted on a bulldozer (blade, rake, V-plow, or teeth), or dragged behind a tractor (Imsett or S.F.I. scarifier, or rolling chopper). Drag type units designed and constructed by Ontario's Department of Lands and Forests used anchor chain or tractor pads separately or in combination, or were finned steel drums or barrels of various sizes and used in sets alone or combined with tractor pad or anchor chain units.

J. Hall's (1970) report on the state of site preparation in Ontario noted that blades and rakes were found to be well suited to post-cut scarification in tolerant hardwood stands for natural regeneration of yellow birch. Plows were most effective for treating dense brush prior to planting, often in conjunction with a planting machine. Scarifying teeth, e.g., Young's teeth, were sometimes used to prepare sites for planting, but their most effective use was found to be preparing sites for seeding, particularly in backlog areas carrying light brush and dense herbaceous growth. Rolling choppers found application in treating heavy brush but could be used only on stone-free soils. Finned drums were commonly used on jack pine–spruce cutovers on fresh brushy sites with a deep duff layer and heavy slash, and they needed to be teamed with a tractor pad unit to secure good distribution of the slash. The S.F.I. scarifier, after strengthening, had been "quite successful" for 2 years, promising trials were under way with the cone scarifier and barrel ring scarifier, and development had begun on a new flail scarifier for use on sites with shallow, rocky soils. Recognition of the need to become more effective and efficient in site preparation led the Ontario Department of Lands and Forests to adopt the policy of seeking and obtaining for field testing new equipment from Scandinavia and elsewhere that seemed to hold promise for Ontario conditions, primarily in the north. Thus, testing was begun of the Brackekultivator from Sweden and the Vako-Visko rotary furrower from Finland.

Mounding

Site preparation treatments that create raised planting spots have commonly improved outplant performance on sites subject to low soil temperature and excess soil moisture. Mounding can certainly have a big influence on soil temperature. Draper et al. (1985), for instance, documented this as well as the effect it had on root growth of outplants (Table 30).

The mounds warmed up quickest, and at soil depths of 0.5 cm and 10 cm averaged 10 and 7 °C higher, respectively, than in the control. On sunny days, daytime surface temperature maxima on the mound and organic mat reached 25 °C to 60 °C, depending on soil wetness and shading. Mounds reached mean soil temperatures of 10 °C at 10 cm depth 5 days after planting, but the control did not reach that temperature until 58 days after planting. During the first growing season, mounds had 3 times as many days with a mean soil temperature greater than 10 °C than did the control microsites.

Draper et al.'s (1985) mounds received 5 times the amount of photosynthetically active radiation (PAR) summed over all sampled microsites throughout the first growing season; the control treatment consistently received about 14% of daily background PAR, while mounds received over 70%. By November, fall frosts had reduced shading, eliminating the differential. Quite apart from its effect on temperature, incident radiation is also important photosynthetically. The average control microsite was exposed to levels of light above the compensation point for only 3 hours, i.e., one-quarter of the daily light period, whereas mounds received light above the compensation point for 11 hours, i.e., 86% of the same daily period. Assuming that incident light in the 100-600 $\mu Em^{-2}s^{-1}$ intensity range is the most important for photosynthesis, the mounds received over 4 times the total daily light energy that reached the control microsites.

Orientation of Linear Site Preparation, e.g., Disk-Trenching

With linear site preparation, orientation is sometimes dictated by topography or other considerations, but the orientation can often be chosen. It can make a difference. A disk-trenching experiment in the Sub-boreal Spruce Zone in interior British Columbia investigated the effect on growth of young outplants (lodgepole pine) in 13 microsite planting positions: berm, hinge, and trench in each of north, south, east, and west aspects, as well as in untreated locations between the furrows. Tenth-year stem volumes of trees on south, east, and west-facing microsites were significantly greater than those of trees on north-facing and untreated microsites. However, planting spot selection was seen to be more important overall than trench orientation.

In a Minnesota study, the N–S strips accumulated more snow but snow melted faster than on E–W strips in the first year after felling. Snow-melt was faster on strips near the centre of the strip-felled area than on border strips adjoining the intact stand. The strips, 50 feet (15.24 m) wide, alternating with uncut strips 16 feet (4.88 m) wide, were felled in a *Pinus resinosa* stand, aged 90 to 100 years.

TREE MEASUREMENT

Trees have a wide variety of sizes and shapes and growth habits. Specimens may grow as individual trunks, multitrunk masses, coppices, clonal colonies, or even more exotic tree complexes. Most champion tree programs focus finding and measuring the largest single-trunk example of each species. There are three basic parameters commonly measured to characterize the size of a single trunk tree: height, girth, and crown spread. Additional details on the methodology of Tree height measurement, Tree girth measurement, Tree crown measurement, and Tree volume measurement are presented in the links herein. A detailed guideline to these basic measurements is provided in The Tree Measuring Guidelines of the Eastern Native Tree Society by Will Blozan.

Summaries of how to measure trees are also presented by various groups involved in documenting big trees around the world. These include among others: a) American Forests Tree Measuring Guidelines; b) National Register of Big Trees - Australia's Champion Trees: Tree Measurement, Champions and Verification; c)Tree Register: A unique record of Notable and Ancient Trees in Britain and Ireland - How to measure trees for inclusion in the Tree Register; and d) NZ Notable Trees Trust.Other parameters also measured include trunk and branch volume, canopy structure, canopy volume, and overall tree shape. Overviews of some of these more advanced measurements are discussed in Blozan above and in "Tsuga Search Measurement Protocols" by Will Blozan and Jess Riddle, September 2006, and tree trunk modeling by Robert Leverettand Leverett and others. The appropriate measurement protocols for multitrunk trees and other more exotic forms are less well-defined, but some general guidelines are presented below.

Height

Tree height is the vertical distance between the base of the tree and the highest sprig at the top of the tree. The base of the tree is measured for both height and girth as being the elevation at which the pith of the tree intersects the ground surface beneath, or "where the acorn sprouted." On a slope this is considered as halfway between the ground level at the upper and lower sides of the tree. Tree height can be measured in a number of ways with varying degrees of accuracy. There are direct measurements. Shorter trees can be measured using a long pole extended vertically to the top of the tree. Larger trees can be climbed and a tape measurement made from the highest point of the climb to the base of the tree. The distance to the top of the tree can be measured from that point, if needed, using a pole. Historically the most direct method of finding a tree's height was to cut it down and measure it prostrate on the ground.

Tree heights can also be measured remotely from the ground. The most basic remote height methodologies are all variations of the stick measurement. The height is calculated using the principle of similar triangles. A short stick is held out pointing vertically at arm's length by its base pointing vertically. The surveyor moves in and out toward a tree until the base of the stick above the lower hand aligns with the base of the tree and the top of the stick aligns with the top of the tree. The distance from the lower hand to the surveyor's eye is measured, the distance from the lower hand to the top of the stick is measured, and the distance from the eye to the base of the tree is measured with a tape. The ratio of distance from the eye to the hand is to the distance from eye to the base of the tree, as is equal to the ratio of the length of the stick to the height of the tree provided that the top of the tree is positioned vertically over the base.

(distance from eye to base of tree/distance from eye to base of stick) x length of stick = tree height

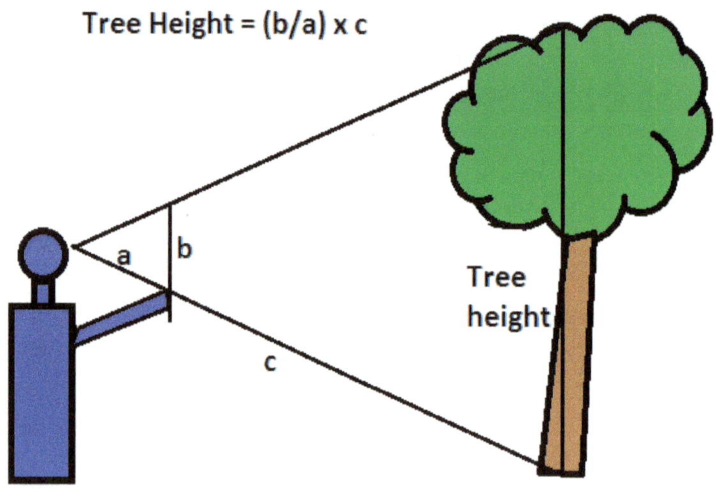

Stick measurement method.

A second method uses a clinometer and a tape and is commonly employed in the forestry industry. This process applies the trigonometric tangent function. In this process the horizontal distance is measured to the trunk of a tree from a sighting position. The angle to the top of the tree is measured with the clinometer.

Horizontal distance at eye level to the tree trunk x tangent Θ = height above eye level

If the tree extends below eye level the same process is used to determine the length below eye level and that is added to height above eye level to determine total tree height. Different clinometers have different reading scales, but all apply the same function. Calculations are as above if it reads in degrees. If it has a percentage scale, then the percentage is multiplied by the distance to the tree to determine a height or extension above and below eye level. Others have a 66' scale where if used at a distance of 66 feet from the tree the height above or below eye level can be directly read from the scale. Errors frequently accompany this type of measuring. The process assumes that the top of the tree is directly over the base of the tree. The tree top may be offset significantly from a point directly over the base (or level point on the trunk). Typical errors from this effect are often on the order of 10 to 20 feet. A bigger problem is misidentifying a forward leaning branch for the actual top of the tree. Errors associated with this mistake may yield measurements that are off many tens of feet and some errors of over forty feet and more have made it onto various champion tree lists, and in at least two cases, the errors have exceeded 60 feet.

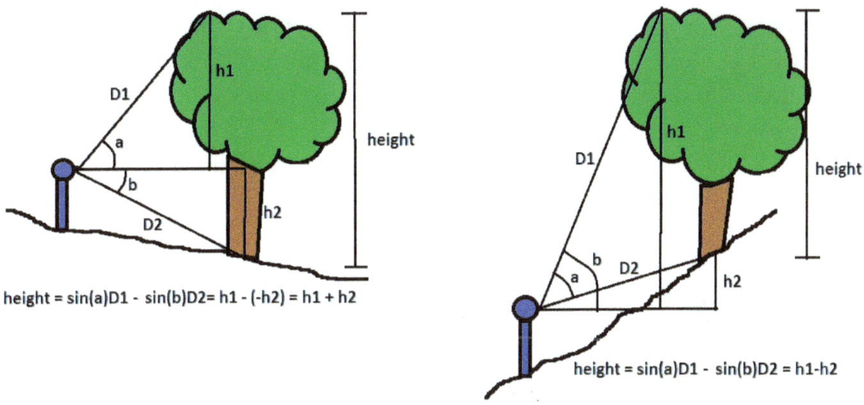

Sine height measurement.

A third remote method is called the sine-height method or ENTS method. It requires the use of a laser rangefinder and clinometer. In this method the distance to the top of a tree is directly measured by the surveyor using the laser rangefinder. The angle to the top is measured with the clinometer. The height of the tree top above eye level: [height = sine Θ of angle x distance to the top] the same process is used to measure the extension of the base of the tree below or even above eye level. Since the measurement is along the hypotenuse of the right triangle and both the top and bottom triangles are independent, it doesn't matter if the tree top is offset from the base as this does not affect the calculation. In addition the top branches of the tree can be scanned with the laser rangefinder to find which top is actually the tallest and the major error of misidentifying the top can be avoided. If the true top of the tree is misidentified the height measurement for the tree will simply be short by some amount, and will not be exaggerated. The height will be correct for the target being measured. With calibration, multiple shots, and a technique to deal with scales that read only to the nearest yard or meter, tree heights can typically be measured to within one foot using this methodology. Other measurement techniques include surveying with a transit and a total station, extended baseline method, the parallax method, and the three verticals method.

Girth

Girth is a measurement of the distance around the trunk of a tree measured perpendicular to the axis of the trunk. Use of girth to arrive at an equivalent diameter is an older forestry measurement that is still used. In the United States girth is measured at a height of 4.5 feet above ground level. Elsewhere in the world it is measured at a height of 1.3 meters,1.4 meters, or 1.5 meters.

Tree girth measurement is commonly performed by wrapping a tape around the trunk at the correct height. Tree girth may also be measured remotely using a monocular w/ reticle, through photographic interpretation, or by some electronic surveying instruments. In these remote methods a diameter perpendicular to the surveyor is what is actually being measured and that is converted to girth by multiplying that number by pi. Many trees flare outward at their base. The standard height up the trunk for determining girth is easily measurable and in most trees it is above the majority of the basal flare and gives a fair approximation of the size of the trunk. For the largest trees, or those with a broad basal flare far up the trunk, it would be appropriate to also measure a second girth above the flare and to note this height.

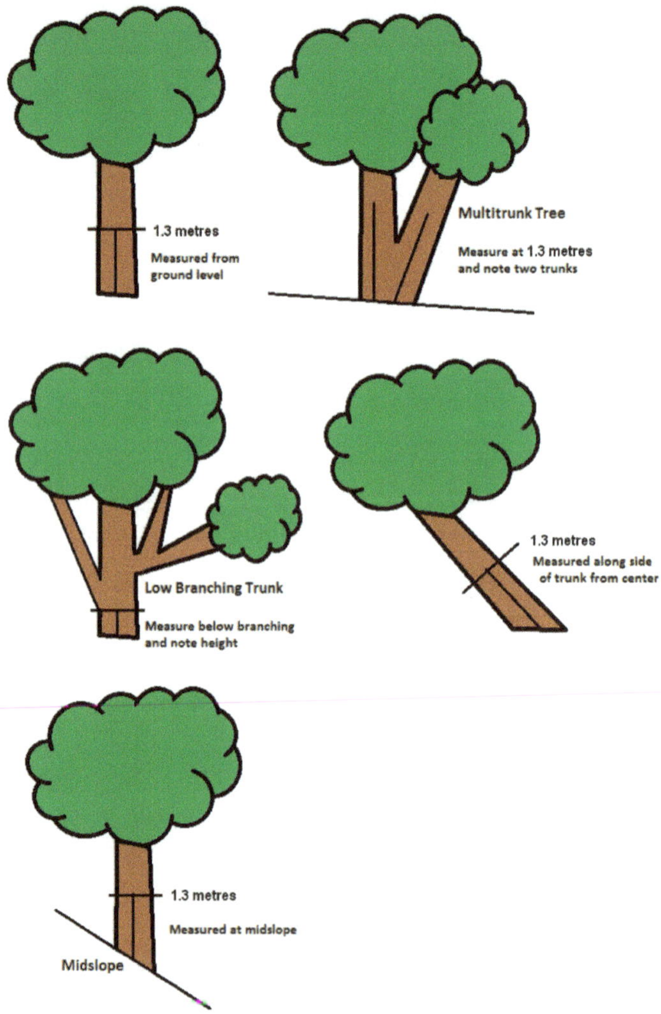

Tree girth measurement diagram.

If there are significant low branches that emerge below this height, ignoring any minor epicormic sprouts and dead branches, then the girth should be measured at the narrowest point below the lowest branch and that height noted. If there is a burl or protuberance at the measurement height, then the girth should be measured immediately above the protuberance or at the narrowest point of the trunk below the protuberance and that height noted.

If the tree is growing on sloping ground the base of the tree should be considered to be where the center or pith of the tree intersects the supporting surface below, typically at mid-slope along the side of the tree. If the tree is large and this measurement would place one portion of the girth loop below ground level, then the measurement should be made at 4.5 feet above ground level at the high side of the slope.

It should always be noted whether or not the tree being measured is single or multitrunk. A single trunk tree is one which would only have a single pith at ground level. If the tree would have more than one pith at ground level, it should be listed as a multitrunk tree and the number of trunks included in the girth measurement noted. If the flare at the base of the tree extends above this default girth height, then ideally a second girth measurement should be collected where possible above the basal flare and this height noted.

Crown Spread

Crown spread is a measure of the footprint or plan area of the crown of the tree expressed as a diameter. The most basic crown spread measurement is the average length of two lines across the crown area. The first measurement is made along the longest axis of the crown from one edge to the opposite edge. A second measurement is taken perpendicular to the first line through the central mass of the crown. The two values are averaged to calculate crown spread. A second method is to take a series of four or more spokes running from the side of the center of the trunk to the edge of the crown. The more spokes measured the better the crown size is represented. The crown spread is the twice the average of all of the spokes. For trees in open areas, crown spreads can also be measured where high enough resolution photos are available using Google Earth. There are length measurements tools built into the program that will allow multiple diameters across the crown to be measured or averaged. Alternatively there are add-on software packages available that can allow an area to be outlined on the Google Earth image and the area enclosed calculated. This can then be converted to crown spread. Leverett has also provided four options for measuring the crown area through compass and clinometer surveys around the outer edge of the crown or through a combination of measurements from the edge of the crown and to the trunk, and those around the crown perimeter. Maximum crown spread and maximum limb length measurements can also be measured if needed.

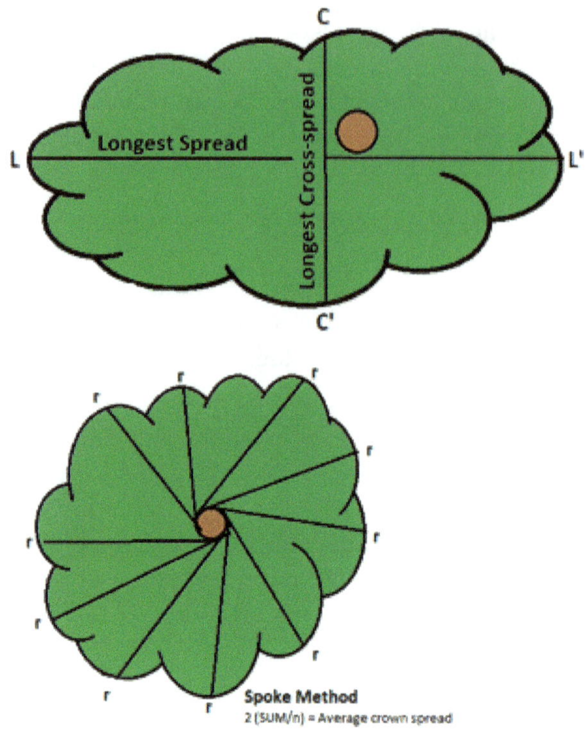

Tree crown spread measurement.

Crown Volumes can be measured as an extension of the basic crown spread measurement. One method is to map a network of points on the outer surface of the crown from various positions around the tree and plotted by map position and height. The crown itself can be subdivided into smaller segments and the volume of each segment calculated individually. For example, the crown could be subdivided into a series of disk-like slices by elevation, the volume of each disk calculated, and the total of all the disks are added together to determine crown volume. Frank developed a simpler method that requires measuring average crown spread, height of the crown from base to top, and matching the crown profile to best fit of a family of crown profile shapes. The method calculates the volume enclosed by rotating the selected profile around the axis of the tree given the measured crown length and average crown spread.

Volume and Canopy Mapping

Tree volume measurements may include just the volume of the trunk, or may also include the volume of branches. Volume measurements can be achieved via ground based or aerial methods. Ground based measurements are obtained by the use of a monocular w/ reticle, laser rangefinder, and a clinometer. Aerial measurements are direct tape measures obtained by a climber in the tree. A monocular w/reticle is a small telescope with an internal scale visible through the glass. The monocular is mounted on a tripod and the trunk of the tree is sighted through the lens. The width of the trunk is measured as so many units of the reticle scale. The height and distance of the target point is measured using the laser rangefinder and clinometer. With the distance known, the diameter of the tree measured expressed as units of the reticle scale, and an optical scaling factor for the monocular w/reticle, provided by the manufacturer and calibrated by the user, the diameter of the tree at that point can be calculated:

Diameter= (Reticle scale) X (distance to target) ÷ (optical factor)

A series of tree diameters up the trunk of the tree are systematically measured using this procedure from the base of the tree to the top and their height is noted. Some photographic methods are being developed to allow calculation of diameters of trunk and limb segments in photographs that contain a scale of known size and where distance to the target is known.

Tree climbers can physically measure the circumference of the tree using a tape. The climber will ascend into the tree until he reaches the highest safe climbing point. Once this point is reached a tape is pulled up along the side of the trunk via a drop rope. The upper end of the tape is lightly tacked at this point and allowed to hang freely down the trunk. The distance from the highest climb point and the top of the tree is measured using a pole that extends from the tree top to the anchor point of the tape. This height is noted and the diameter of the tree is measured at that point. The climber then rappels down the tree measuring the trunk circumference by tape wrap at different heights with the height of each measurement referenced to the fixed tape running down the trunk. Whether using the aerial or ground based methods, the diameter or girth measurements do not need to be evenly spaced along the trunk of the tree, but a sufficient number of measurements need to be taken to adequately represent the changes in diameter of the trunk.

To calculate trunk volume, the tree is subdivided into a series of segments with the successive diameters being the bottom and top of each segment and its length equal to the difference in height between the lower and upper diameter. Cumulative trunk volume is calculated by adding the volume of the measured segments of the tree together. The volume of each segment is calculated as the volume of a frustum of a cone where:

$$\text{Volume} = h(\pi/3)(r_1^2 + r_2^2 + r_1 r_2)$$

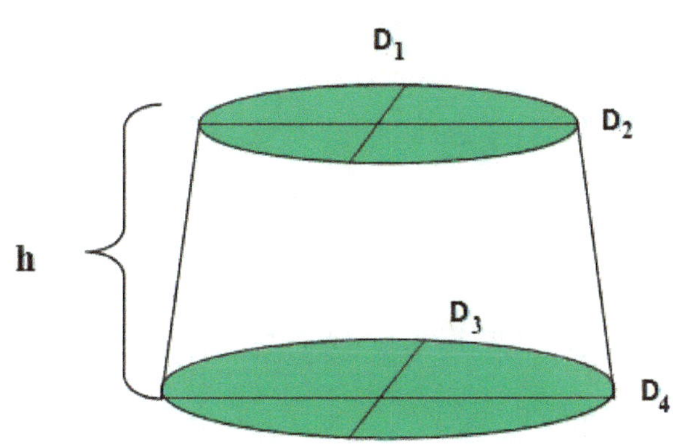

Frustum of a cone.

A similar, but more complex formula can be used where the trunk is significantly more elliptical in shape where the lengths of the major and minor axis of the ellipse are measured at the top and bottom of each segment.

In areas where the trunk bifurcates the trunk will not have a circular or simple elliptical diameter. Blozan as part of the Tsuga Search Project created a wooden frame that would fit around the odd shaped section and measured the position of the surface of the trunk with respect to the frame. These points were plotted on a graph and the cross-sectional area of the irregular trunk shape was calculated. This area was, in turn, converted to an equivalent circular area for use in the volume formula.

Many trees flare outward significantly at the base and this basal wedge has a complex surface of bumps and hollows. This becomes an even more complex volume in trees growing on a slope. Approximations of the volume of this basal segment using best estimates of the effective diameters exhibited may be used in many cases. In other cases footprint mapping is an option. In footprint mapping a level, rectangular reference frame is placed around the base of the tree to create a horizontal plane. The position of the multiple points on the trunk surface is measured with respect to the frame and plotted. This process is repeated at different heights creating a series of virtual slices at different heights. The volume of each individual slice is then calculated and all are added together to determine the volume of the basal wedge. Taylor has been developing a cloud mapping process using optical parallax scanning technology whereby thousands of measurements are made around the trunk of a tree. These can be used to recreate a three-dimensional model of the trunk and volume data is among the values that can be calculated.

Canopy mapping is the process whereby the positions and size of the branches within the canopy are mapped in three-dimensional space. It is a labor-intensive process that is usually reserved for only the most significant specimens. This is usually done from a set position or a series of positions within the tree. Sketches and photographs are used to facilitate the process. Trees are climbed and the overall architecture is mapped including the location of the main stem and all reiterated trunks, in addition to all branches that originate from trunks. The position of every branch point in the canopy down to a certain size and also the positions of various reiterations, breaks, kinks, or any other eccentricities in the tree are also mapped. Each mapped trunk and branch is measured for basal diameter, length, and azimuth. Climbers measure specific circumferences and detail other features within the tree

Trees with Unusual Forms

Not all trees have a single trunk, and others pose additional measurement problems because of their size or configuration. The odd forms include those forms that grew because of unusual circumstances that affected the tree, or those trees that simply have an unusual growth form not seen in most other tree species. Frank proposed a classification system for various tree forms: 1) Single Trunk Trees; 2) Multitrunk Trees; 3) Clonal Coppices; 4) Clonal Colonies; 5) Conjoined and Hugging Trees; 6) Fallen Trees; 7) Tree complexes, and 8) Banyan-like trees; 9) Trees with Large Aerial Root Systems; and 10) Epiphytic Trees. This initial framework has continued to evolve in discussions within the NTS, but provides an initial beginning and suggestions on how to approach measuring these various tree growth forms.

Since most of these trees are unique or unusual in their form and not amenable to easy measurement, the recommended approach is to write a detailed narrative description of the tree with what measurements that can be taken to amplify and better illuminate the descriptions. These trees should be documented even if the results are in the form of a written narrative rather than a collection of numerical measurements.

There are some parameters that should be consistently measured whenever possible, height is one example. The cross-sectional areas occupied by the trunks and the crown are also parameters that are generally measurable. Other measurements could be taken where they seem to add to the narrative description of that particular tree. GPS locations should be taken whenever possible. Absent a GPS instrument, the locations should be pulled from Google Maps, or topographic maps. Beyond these basics, values like number of trunks larger than a prescribed value, the maximum girth of the largest trunk, and whatever seems appropriate for that particular tree grouping should be recorded. Photographs of these unusual trees are important as they can immensely improve the understanding of what is being described, and help others to visualize the tree. A process or system is needed whereby the photos of a particular tree can be associated with the description of the tree in the researcher's notes. The goal of the narrative and measurements is to document the tree or tree grouping.

Single trunk trees can also pose measurement problems. Consider trees with very large girths, such as some of the sequoias growing the western United States. If they are growing on even a gentle slope, if girth is measured at 4.5 feet above where the pith of the tree emerges from the ground, the upside of the tape could easily be below ground level. In this case a better option would be to measure the standard girth measurement at 4.5 feet above ground level on the high side of the tree and note this in the measurement description. If measuring a mountaintop forest of stunted trees only six feet tall, a girth measurement made at 4.5 foot would be meaningless. In the case of these stunted trees a girth taken at 1 foot above the base might be more appropriate. Girth measurements should be taken at the standard heights whenever possible. Where this measurement is not meaningful, an additional girth measurement should be taken at a more appropriate position and that height noted.

A double-trunked cherry tree.

Multitrunk trees are the most common form after single trunk trees. Often these represent separate trunks growing from a single root mass. This occurs frequently in some species when the initial trunk has been damaged or broken and in its place two or more new shoots grow from the original root mass. These are genetically the same, but as their growth form is different they should be considered as a different measurement category than single trunk trees. These multiple trunks commonly will grow together to form a large combined mass at the base and split into individual trunk at greater heights. If they are individual trunks at breast height then the individual trunks can be measured separately and treated as individual single trunk trees. If they have grown together at breast height then a measurement of their combined girth should be made at that height the number of trunks incorporated into the girth measurement listed. If the tree splays outward dramatically at breast height, then the girth should be measured at the narrowest point between breast height and the ground and that height noted. Other girth measurements guidelines outlined for single trunk trees, such as low branches and burls, apply to multitrunk girths as well. The height of the tallest trunk in the multitrunk specimen would then be the height of the multitrunk specimen and the combined crown spread of all the individual trunks the multitrunk specimen collectively would be the multitrunk crown spread. If one of the individual trunks is significantly larger than all the others, it can be treated as if it were a single trunk tree. Its girth is measured where it emerges from the combined mass, and the height and crown spread of that particular trunk is measured individually.

Clonal colonies, such as the Pando aspen, may occupy many acres. The area occupied by the colony should be measured as well as the size of the largest individual trunk present.

Banyan-like trees similarly consist of multiple trunks spread across a large area. In many of these specimens the interior trunks are not easily accessible. An approach to their measurement would be to measure the area occupied by the many trunks, the area occupied by the crown of the tree, the height of the tree, and any other measurements the investigator deems appropriate. These measurements would then be supplemented by a narrative description and photographs. The goal in all of these cases of trees with unusual forms is to document their characteristics.

Tree Shape

Different tree species tend to have different shapes and tree shapes also vary within a single tree species. As a general observation trees growing in an open setting tend to be shorter and have broader crowns, while those growing in a forested setting tend to be taller and have narrower crowns. In forested areas trees grow taller and put more energy into height growth as they compete with other trees for the available light. Often the tallest examples of many species are found where they are a secondary species on a site and competing for light with other taller tree species. The tall bay laurel (Umbellularia californica) at 169.4 feet discovered by Zane Moore in Henry Cowell Redwoods State Park is an example of an exceptionally tall understory tree growing among other taller species.

Ternary Tree Shape Plots. A methodology for plotting different tree shapes graphically was developed by Frank using ternary plot diagrams. Ternary plots can be used to graphically display any set of data that includes three terms which total to some constant.

Generally this constant is 1 or 100%. This is ideal for plotting the three most commonly measured tree dimensions. The first step in the analysis is to determine what an average shape for trees is in general. These three basic parameters can be expressed as a ratio of height to girth to average crown spread. Some trees are tall and narrow, while others are low and broadly spreading. The data used to determine the average tree shape is derived from a tabulation of the largest trees of each of 192 different species in the NTS 2009 dataset. The average girth, height, and crown spread values were calculated for the measurements included in the listing. For the dataset the average height was 87.6 feet, the average girth was 100.1 inches, and the average spread was 54.9 feet. It is not critical that these values be exact for analysis purposes. The next step is to standardize each measured parameter. The quantity measured for a particular tree is divided by the standard value as determined above. The next step is to normalize the data set so that the sum of the three parameters expressed as a percentage will equal one. This enables the shapes of different trees of different sizes to be compared. The final step is to plot these results as a ternary graph to better compare the results. As an example, the measurement data for 140 live oaks measured as part of the NTS Live Oak Project were graphically plotted using this process.

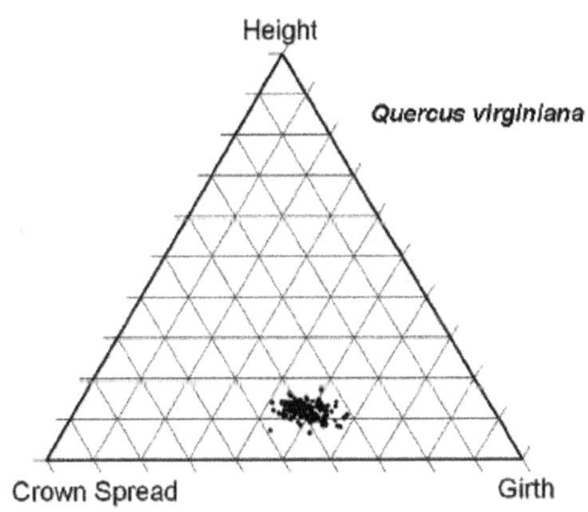

Figure 5. Plot of the normalized data of the 140 live oak trees included in the tree shape analysis.

Tree shape diagram for live oak

The cluster representing the live oak data falls on the extreme edge of the general pattern of tree shapes. The height proportion exhibits a maximum of 17.23% of the shape value and a minimum of 6.55%, the girth (minimum of 19 feet in the data set) exhibits a maximum of 58.25% and a minimum of 40.25%, and Average Crown Spread maximum of 49.08% and a minimum of 30.92%. These points represent the measurements of the largest specimens of live oak measured in the field and typically represent open grown specimens, but the tightness of the shape cluster is still remarkable. It is even more interesting to note that while the data set contains both multiple trunk trees and single trunk trees, both plot within the same tight cluster.

Dendrochronology

Dendrochronology is the science of dating and study of annual rings in trees. A tree in temperate and colder climates typically will grow one new ring every year, therefore the age of the tree can theoretically be determined by counting the number of rings present. The problem lies in the fact that some years, particularly in years of drought, a tree will not grow an annual ring. In other years where the growing season is interrupted a tree may grow a second false ring. The trees rings are commonly measured by taking a series of core samples. A borer is used to extract a pencil-size diameter or smaller core from a living tree or from a log. For downed and dead trees a disk section or "tree cookie" may also be taken, these are polished, the rings identified, and the number of rings and the distance between each are recorded. By comparing rings from multiple trees, through cross-dating, a dendrochronologists can determine if rings are missing or if false rings are present. Through this process the tree ring record can be used to investigate past climatic conditions. Tropical trees often lack annual rings and ages for these trees can be measured using radiocarbon dating of wood samples from the trees.

There are two major listings of the maximum ages of trees. The OldList is a database of ancient trees maintained by Rocky Mountain Tree-Ring research. Its purpose is to identify maximum ages that different species in different localities can attain such that exceptionally old age individuals may be recognized. In addition to the original OldList, Neil Pederson at the Tree Ring Laboratory of Lamont-Doherty Earth Observatory and Columbia University has created an eastern OLDLIST focused on old trees in the eastern North America. In addition to these sources of tree ring data there is the ITRDB. The International Tree-Ring Data Bank is maintained by the NOAA Paleoclimatology Program and World Data Center for Paleoclimatology. The Data Bank includes raw ring width or wood density measurements, and site chronologies. Reconstructed climate parameters, including North American Drought, are also available for some areas. Over 2000 sites on six continents are included.

The oldest known tree is a Great Basin bristlecone pine, Pinus longaeva, growing in the White Mountains of eastern California. The tree was cored by Edmund Schulman in the late 1950s, but was never dated. Recently Tom Harlan completed the dating of the old core sample. The tree is still alive and is 5062 years old as of the 2012 growing season. Older ages are given for sprouts growing from roots or clonal colonies, but these values are not from an individual stem that has persisted for that duration. Whether these are considered to be older trees or not is dependent on the definition of the term 'tree'.

In spite of the extensive amounts of work done by dendrochronologists in investigating trees, the maximum ages attainable by most common species is not clear. Dendrochronologists typically focus on trees that are known to have a long life span when investigating a site. This is because their goal is paleoclimatic reconstruction or archaeological investigation and longer lived trees provide longer data record. Most species believed to be shorter lived have not been systematically investigated and cross dated. The Native Tree Society is compiling basic ring counts for many of these species in order to better understand the age structure of the forests they are investigating, with the recognition that ring counts ages may be off due to missing or false rings.

Big Tree Formulas

American Forest Formula. American Forests has developed a formula for calculation tree points for determining champion trees for each species. Three measurements: Trunk Circumference (inches), Height (feet), and Average Crown Spread (feet). Trees of the same species are compared using the following calculation:

Trunk Circumference (inches) + Height (feet) + ¼ Average Crown Spread (feet) = Total Points.

The American Forests National Big Tree Program is the largest in the world with coordinators in all fifty states and the District of Columbia and is used as a model for several Big Tree programs around the world. American Forests describe it as a conservation movement to locate, appreciate and protect the biggest tree species in the United States with more than 780 champions crowned each year, with 200 more species without a crowned champion in 2012, and documented in their biannual publication — the National Register of Big Trees. The program has been operating since 1940.

For example, the Australian National Register of Big Trees uses the American Forests formula. The individual measurements are listed using both Imperial and metric values. Trees must be single-stemmed at 1.4m above ground where circumference is measured. They write that making the Australian tree Points directly comparable to the U.S.A. is important because Australians can view the American Forests Register of Big Trees and gain immediate, and much enjoyment, in comparing their Champions with ours, and vice versa for North American tree enthusiasts viewing our NRBT.

Tree Dimension Index. The Native Tree Society, in addition to the American Forests formula, uses an alternative approach to compare relative sizes of trees, both within the same species and against others. The Tree Dimension Index (TDI) is highly adaptable and can be tailored to reflect the attributes of an individual tree and how they compare relative to the largest known specimen. The premise is that the specific dimensions of the tree are given a value (percentage) that reflects its relative rank against the maximum known for the same dimension for the species. For example, the tallest known eastern hemlock would get a value of 100 for height since it represents 100% of the maximum value known for the species. A shorter tree that was 75% of the maximum known height would get a value of 75 for its height. Likewise, the values of diameter and volume would be determined by the relative value when ranked against the known maxima. With three ranked attributes the maximum TDI value would theoretically be 300. However, this would represent one tree exhibiting all three maxima- an unlikely possibility. However, the apparent size of a tree can be realized by ranking the cumulative values against the theoretical maximum. A tree scaling close to 300 would suggest that it was nearly the largest specimen theoretically possible based on currently known maxima. A two-value TDI using height and girth was presented for 259 white pines (Pinus strobus) by Friends of Mohawk Trail State Forest to MA DCNR in 2006. The TDI values in the dataset ranged from 172.1 to 125.2 out of a possible maximum of 200.

Location

As with any other scientific investigation it is critical to establish the location of the trees being investigated. Without that information, the location of tree may be lost and other investigators will not be able to relocate the tree in the future. There is also the possibility that the same tree might be misidentified and re-measured as a different tree. GPS locations should be taken for every tree measured. GPS in most cases is accurate enough to distinguish the location of a specific tree. The actual accuracy users attain depends on a number of factors, including atmospheric effects and receiver quality. Real-world data collected by the FAA show that some high-quality GPS SPS receivers currently provide better than 3 meter horizontal accuracy. If GPs is not available then the approximate latitude and longitude data should be topographic maps, or air photo sources such as Bing Maps, Google Earth, or similar services.

Tree Databases

Several of the larger tree groups maintain interactive databases of tree information. Different types of information are available on different databases and there are different requirements for data entry. American Forests provides a searchable database of their Champion trees and in 2012 included data on 780 tree species. Most of the individual state big programs are administrated through the American Forest Big Tree Programs. The Native Tree Society has its own database Trees Database with the requirement that the trees entered meet their height measurement standards. There also are databases maintained by Australia's National Register of Big Trees, Notable trees of New Zealand, Monumental Trees (primarily focused on Europe but including trees from elsewhere in the world), and The Tree Register - A unique record of Notable and Ancient Trees in Britain and Ireland.

There are many other sites maintained by groups and individuals that include tabulations of big trees of a specific area, of a particular species, or simply the largest individuals. Some of these include Landmark Trees, Native Tree Society,baumzaehlen, Old Trees in The Netherlands and Western Europe, the big eucalypts of Tasmania and Victoria, and the Old Growth Forest Network.

In all cases, the data collected needs to be organized into a useful searchable format that can be used. The Native Tree Society provides a free downloadable excel spreadsheet that can be used for organizing tree datasets Tree Measurement Data Spreadsheet. The spreadsheet can be modified for the needs of the user.

Rucker Indexes

Rucker Index is a family of indexes that are used to compare tree population between different tree sites. It is not species dependent and can be applied to sites of different sizes. The basic Rucker Index is a measure of overall tree height. The Rucker Height Index 10 or RI10 is the numerical average of the height in feet of the tallest individual of each of the ten tallest species on a site. A particular species enters the index only once. The index provides a numerical evaluation of both maximum height and diversity of the dominant species. High index values are the result of many factors, including climate, topography, soils, and a lack of disturbance. While the most extensive sites benefit from a greater variety of habitat and more individual trees, some exceptional sites are quite small. The Rucker Height Index is essentially a foreshortened version of a complete profile of all the species found on a particular site.

Variations of the Rucker Index can also be calculated. If the site has a high diversity of species a RI20 can be calculated using twenty species. For sites with limited data or low species diversity, a RI5 with only five species can be calculated. A Rucker Girth Index or RGI10 can also be calculated using the girth of the largest girth individual in each of the ten fattest species on a site.

The Rucker Height Index or Rucker Index has a numerous merits that make it a useful measurement when comparing various tall tree sites: 1) The formula is straight forward, unambiguous, and easy to apply; 2) The index can be applied to forests in any area with any make-up of trees; 3) The index requires a fairly diverse mix of trees in order to generate a high index value; and 4) To get a sufficient diversity of trees of great height requires at least a modest size or larger plot of forest and a reasonably thorough examination to generate a high RI value.

In January 2012 the calculated Rucker Index for the World was: 312.39 feet. The Rucker Index for the west coast of North America, and also all of North America is 297 The RI10 for Great Smoky Mountains National Park is 169.24, the tallest site in eastern United States. For the northeastern United States the RI10 is 152.6, and for the southeast, excluding GSMNP the RI10 is 166.9.

REFERENCES

1. Hawley, R.C. and D.M. Smith. *The Practice of Silviculture.* 6th edition. New York: John Wiley & Sons Inc., 1954. Print.
2. Ford-Robertson, F.C. (Ed.) 1971. Terminology of Forest Science, Technology, Practice and Products. English language version. Soc. Amer. For., Washington DC. 349 p.
3. Köstler, J. 1956. Silviculture. Transl. M.L. Anderson from “Waldbau” (1949). Oliver and Boyd, Edinburgh and London, U.K. 416 p.
4. Jeglum, J.K.; Kershaw, H.M.; Morris, D.M.; Cameron, D.A. 2003. Best forestry practives: A guide for the boreal forest in Ontario. Natural Resources Canada, Can. For. Serv., Sault Ste. Marie ON, co-published by Ont. Min. Nat. Resources, Sault Ste. Marie ON. 110 p.
5. John D. Matthews, Silvicultural Systems. Oxford Science Publications. 1991, ISBN 9780198546702
6. Invented by Reiniger at The Schlägl Monastery, Austria http://www.stift-schlaegl.at/prodon.asp?peco=&Seite=373&Lg=1&Cy=1
7. Silviculture Concepts and Applications, Ralph D. Nyland 2002 pg. 20

8. Grossnickle, S.C. 2000. Ecophysiology of northern spruce species: the performance of planted seedlings. NRC Research Press, Ottawa ON. 409 p.
9. 11 Huss,J. 2004;
10. Schopmeyer, C.S. (Tech. Coord.) 1974. Seeds of Woody Plants in the United States. USDA, For. Serv., Washington DC, Agric. Handb. 450. 883 p.
11. MacArthur, J.D.; Fraser, J.W. 1963. Low temperature germination of some eastern Canadian tree seed. For. Chron. 39(4):478–479.
12. Arnott, J.T. 1974. Germination and seedling establishment. p. 55–66 *in* Cayford, J.H. (Ed.). Direct Seeding Symposium, Timmins ON, Sept. 1973, Can. Dep. Environ., Can. For. Serv., Ottawa ON, Proc., Publ. 1339. Cite error: Invalid <ref> tag; name “arn” defined multiple times with different content (see the help page).
13. Alexander, R.R. 1984. Natural regeneration of Engelmann spruce after clearcutting in the central Rocky Mountains in relation to environmental factors. USDA, For. Serv., Rocky Mountain For. Range Exp. Sta., Fort Collins CO, Res. Pap. RM-254. 17 p.
14. Baldwin, H.I. 1927. A humus study in Norway. Ecology 8:380–383.
15. Mork, E. 1933. [Temperature as a factor of regeneration in the spruce forests of Northern Trondhjem.] Meddelelser fra Det Norske Skogforsøkvesen No. 16 Vol. V, No. 1. [Reviewed in J. For. 32:1024, 1934]
16. Rowe, J.S. 1955. Factors influencing white spruce reproduction in Manitoba and Saskatchewan. Can. Dep. Northern Affairs and National Resources, For. Branch, For. Res. Div., Ottawa ON, Project MS-135, Silv. Tech. Note 3. 27 p. Cite error: Invalid <ref> tag; name “rowe” defined multiple times with different content (see the help page).
17. Phelps, V.H. 1940a. Cut-over and burned-over white spruce lands, northern Saskatchewan survey, 1938. Can. Dep. Mines and Resources, Lands, Parks and For. Branch, Dominion For. Serv., Ottawa ON, unpubl. Report. (Cited in Rowe 1955, orig. not seen)
18. Phelps, V.H. 1940b. Spruce regeneration in Canada -the Prairie Provinces. For. Chron. 16:30–37.
19. Alexander, R.R. 1958. Silvical characteristics of Engelmann spruce. USDA, For. Serv., Rocky Mountain For. Range Exp. Sta., Fort Collins CO, Paper 31. 20 p.
20. Sutton, R.F. 1991. Soil properties and root development in forest trees: a review. For. Can., Ont. Region, Sault Ste. Marie ON, Inf. Rep. O-X-413. 42 p.
21. Alexander, R.R. 1983. Seed: seedling ratios of Engelmann spruce after clearcutting in the central Rocky Mountains. USDa For. Serv., Rocky Mountain Forest and Range Exp. Sta., Fort Collins, CO. Research Note RM-426. 6 p. Cite error: Invalid <ref> tag; name “alex3” defined multiple times with different content (see the help page). Cite error: Invalid <ref>tag; name “alex3” defined multiple times with different content (see the help page).

22. Baldwin, H.I. 1942. Forest Tree Seed. Chronica Botanica, Waltham MA. 240 p.

23. Baldwin, H.I. 1933. The density of spruce and fir reproduction related to the direction of exposure. Ecology 14:152–156.

24. Alarik, A. 1925. Moderna huggningsformer tillämpade pâ Finspång. Skogen 12:211–243.

25. Hartley, C. 1918. Stem lesions caused by excessive heat. J. Agric. Res. 14:595–604.

26. Eis, S.; Inkster, J. 1972. White spruce cone production and prediction of cone crops. Can. J. For. Res.2:460–466.

27. Noble, D.L.; Ronco, F. 1978. Seedfall and establishment of Engelmann spruce and subalpine fir in clearcut openings in Colorado. USDA, For. Serv., Rocky Mountain For. Range Exp. Sta., Res. Pap. RM-200. 12 p.

28. Radvanyi, A. 1970. Small mammals and regeneration of white spruce forests in western Alberta. Ecology 51(6):1102–1105.

29. Coates, K.D.; Haeussler, S.; Lindeburgh, S.; Pojar, R.; Stock, A.J. 1994. Ecology and silviculture of interior spruce in British Columbia. Canada/British Columbia Partnership Agreement For. Resour. Devel., Victoria BC, FRDA Rep. 220. 182 p.

30. Gilmour, J.G.; Konishi, J. 1965. Scarification in the spruce alpine fir type of the Prince George Forest District. Preliminary evaluation of methods and resulting regeneration.B.C. For. Serv., Victoria BC, For. Manage. Notes 4. (Cited in Coates et al. 1994).

31. Ackerman, R.F. 1957. The effect of various seedbed treatments on the germination and survival of white spruce and lodgepole pine seedlings. Can. Dep. Northern Affairs & National Resour., For. Branch, Ottawa ON, Tech. Note 63. 23 p.

32. Zasada, J.C. 1985. Production, dispersal and germination, and first-year seedling survival of white spruce and birch in the Rosie Creek burn. p.34–37 *in* Juday, G.P.; Dyrness, C.T. (Eds.), Early Results of the Rosie Creek Fire Research Project, 1984. Univ. Alaska, Agric. For. Exp. Sta., Fairbanks AK, Misc. Publ. 85–2.

33. Zasada, J. 1986. Natural regeneration of trees and tall shrubs on forest sites in interior Alaska. p. 44–73 *in* Van Cleve, K.; Chapin, F.S.; Flanagan, P.W.; Viereck, L.A.; Dyrness, C.T. (Eds.). Forest Ecosystems in the Alaskan Taiga: a Synthesis of Structure and Function. Springer-Verlag New York NY.

34. Zasada, J.C.; Norum, R. 1986. Prescribed burning white spruce slash in interior Alaska. North. J. Appl. For. 3(1):16–18.

35. Bell, F.W. 1991. Critical silvics of conifer crop species and selected competitive vegetation in northwestern Ontario. For. Can., Sault Ste. Marie, Ont./Ont. Min. Nat. Resour., Northwestern Ont. For. Tech. Devel. Unit, Thunder Bay ON, COFRDA Rep. 3310/ NWOFTDU Tech. Rep. 19. 177 p.

36. Anon. 1961. Northern conifer planting stock produced under long photoperiods in Florida. p. 9–13 *in* 1960 Rep., Lake States For. Exp. Sta.

37. *Nienstaedt, Hans; Zasada, John C. (1990). "Picea glauca". In Burns, Russell M.; Honkala, Barbara H. Conifers. Silvics of North America. Washington, D.C.:*

United States Forest Service (USFS), United States Department of Agriculture (USDA). 1 – via Northeastern Area State and Private Forestry (www.na.fs.fed.us).

38. Pollard, D.F.W.; Logan, K.T. 1976. Prescription for the aerial environment for a plastic greenhouse nursery. p.181–191 *in* Proc. 12th Lake States For. Tree Improv. Conf. 1975. USDA, For. Serv., North Central For. Exp. Sta., St. Paul MN, Gen. Tech. Rep. NC-26.
39. Carlson, L.W. 1979. Guidelines for rearing containerized conifer seedlings in the prairie provinces. Can. Dep. Environ., Can. For. Serv., Edmonton AB, Inf. Rep. NOR-X-214. 62 p. (Cited in Nienstaedt and Zasada 1990).
40. Tinus, R.W.; McDonald, S.E. 1979. How to grow tree seedlings in containers in greenhouses. USDA, For. Serv., Rocky Mountain For. Range Exp. Sta., Fort Collins CO, Gen. Tech. Rep. RM-60. 256 p. (Cited in Nienstaedt and Zasada 1990).
41. Tinus, R.W. 1984. Optimum temperatures for growth of southern Rocky Mountain Engelmann spruce and Douglas-fir seedlings. USDA, For. Serv., Rocky Mount. For. Range Exp. Sta., Fort Collins CO, Res. Note RM-442. 5 p.
42. Tappeiner, J.C., D.A. Maguire, and T.B. Harrington. *Silviculture and Ecology of Western U.S. Forests*. Corvallis: Oregon State University Press, 2007. Print.
43. Joyce, D.; Nitschke, P.; Mosseler, A. 2001. Genetic resource management. p. 141–154 *in* Wagner, R.G.; Colombo, S.J. (Eds.). Regenerating the Canadian Forest: Principles and Practice for Ontario. Fitzhenry and Whiteside, Markham ON.
44. Hills, G.A. 1952. The classification and evaluation of site for forestry. Ont. Dep. Lands For., Res. Div., Toronto ON, Res. Rep. 24. 41 p.
45. Mackey, B.G.; McKenney, D.W.; Yang, Y.Q.; McMahon, J.P., and Hutchinson, M.F. 1996. Site regions revisited: a climatic analysis of Hills' site regions for the province of Ontario using a parametric method. Can. J. For. Res. 26:333–354.
46. [OMNR] Ontario Ministry of Natural Resources. 1997. Seed zones of Ontario. Toronto ON, Policy Directive No. FOR 03 02 06 500 02. 5 p. [Cited by Joyce et al. 2005, original not seen]
47. Li, P.; Beaulieu, J.; Bousquet, J. 1997. Genetic structure and patterns of genetic variation among populations in eastern white spruce (*Picea glauca*). Can. J. For. Res. 27:189–198.
48. Toumey, J.W.; Korstian, C.F. 1954. Seeding and Planting in the Practice of Forestry., 3rd ed. Wiley & Son, New York NY. 520 p.
49. Wang, B.S.P. 1973. Collecting, processing and storing tree seed for research use. IUFRO International Symposium on Seed Processing, Bergen, Norway, 1973: Vol. 1:17. 11 p.
50. Noland, T.L.; Mohammed, G.H.; Seymour, N. 2001. Testing tree seed viability with FDA: what does it tell you? p. 23 *in* Proc. Working together for our common future. LUSTR Co-op. Ann. Gen. Meet. & Workshop, Feb. 2001, Timmins ON.
51. Timonin, M.I. 1966. Effect of ultra-sound on the germination of white spruce and jack pine. Can. J. Bot. 44:113–115. (Cited in Coates et al. 1994).

52. Toumey, J.W.; Stevens, C.L. 1928. The testing of coniferous tree seeds at the School of Forestry, Yale University, 1906–1926. Yale Univ. School For., New Haven CT, For. Bull. 21. 46 p.

53. Rafn, J. 1915. The testing of forest seeds during 25 years, 1887–1912. Copenhagen, Denmark. 91 p. (Cited in Toumey and Korstian 1954, orig. not seen)

54. Mittal, R.K.; Wang, B.S.P. 1987. Fungi associated with seeds of eastern white pine and white spruce during cone processing and seed extraction. Can. J. For. Res. 19(9):1026–1034.

55. Haddon, B.D.; Winston, D.A. 1982. Germination after two years storage of artificially ripened white spruce seed. p. 75–80 *in* Wang, B.S.P.; Pitel, J.A. (Eds.). Proc. Internat. Sympos. Forest Tree Seed Storage, Sept. 1980, Chalk River ON. Environ. Can., Can. For. Serv. (Cited in Coates et al. 1994).

56. Wang, B.S.P. 1974a. Testing and treatment of Canadian white spruce seed to overcome dormancy. Assoc. Official Seed Analysts Proc. 64:72–79.

57. Zasada, J.C.; Foote, M.J.; Deneke, F.J.; Parkerson, R.H. 1978. Case history of an excellent white spruce cone and seed crop in interior Alaska: cone and seed production, germination and seedling survival. USDA, For. Serv., Pacific NW For. Range Exp. Sta., Portland OR, Gen. Tech. Rep. PNW-65. 53 p.

58. Armson, K.A.; Sadreika, V. 1979. Forest tree nursery soil management and related practices – metric edition. Ont. Min. Nat. Resour., Div. For. For. Manage. Branch, Toronto ON. 179 p.

59. Simpson, J.D.; Wang, B.S.P.; Daigle, B.I. 2004. Long-term storage of various Canadian hardwoods and conifers. Seed Sci. & Technol. 32:561–572.

60. Heit, C.E. 1961. Laboratory determination and recommended testing methods for 16 spruce (*Picea*) species. p. 165–171 *in* Assoc. Off. Seed Anal. 51st Annu. Meet. Proc. (Cited in Coates et al. 1994).

61. Hellum, A.K. 1968. A case against cold stratification of white spruce seed prior to nursery seeding. Can. Dep. For. and Rural Devel., For. Branch, Ottawa ON, Publ. 1243. 12 p.

62. van den Driessche, R. 1969. Forest nursery handbook. B.C. For. Serv., Victoria BC, Res. Notes 48. 44 p.

63. Santon, J. 1970. Effect of stratification on germination of freshly harvested seed of several spruce and pine species in eastern Canada. Can. Dep. Fish. For., Can. For. Serv., Petawawa For. Exp. Sta., Chalk River ON, Inf. Rep. PS-X-17. 22 p.

64. Wang, B.S.P. 1987. The beneficial effects of stratification on germination of tree seeds. p. 56–75 *in* Proc. Nurserymen's Meeting, Dryden ON, June 15–19, 1987. OMNR, Toronto ON.

65. Hocking, D. 1972. Effects of stratification of Alberta white spruce and lodgepole pine seeds on emergence in operational seedbeds. Environ. Can., Can. For. Serv., Ottawa ON, Bi-mo. Res. Notes 28(4):26–27.

66. Winston, D.A.; Haddon, B.D. 1981. Effects of early cone collection and artificial ripening on white spruce and red pine germination. Can. J. For. Res. 11:817–826.

67. Kolotelo, D. 1997. Anatomy and morphology of conifer tree seed. Forest Nursery Technical Series 1.1 B.C. Min. for., 70 p.

68. Mercier, S. 1991. Maturation et indices de maturité des semences d'épinette blanche. [Maturation and maturity indexes of white spruce seeds.] Quebec Min. For., Quebec QC, Memoire Rech. For. 103. 62 p. [Fr.e.; E 4085]

69. Caron, G.E.; Wang, B.S.P.; Schooley, H.O. 1990. Effect of tree spacing, cone storage, and prechilling on germination of *Picea glauca* seed. For. Chron. 66(4):388–392.

70. Caron, G.E.; Wang, B.S.P.; Schooley, H.O. 1993. Variation in *Picea glauca* seed germination associated with the year of cone collection. Can. J. For. Res. 23(7):1306–1313.

71. ^ Jump up to:[a] [b] Ontario Class Environmental Assessment for Timber Management. 1989. Ont. Min. Nat. Resour. Witness Statements, Evidence Panels I to XVII. Compiled 1989.02,22, Sections separately paginated. **Cite error: Invalid<ref> tag; name "ont" defined multiple times with different content (see the help page).**

72. British Columbia. Ministry of Forests. 1995. Forest practices code of British Columbia: Silvicultural systems guidebook. BC MoF, BC Ministry of Environment, 44 p.

73. Milthorpe, F.L. 1961. The nature and analysis of competition between plants of different species. p. 330–355 *in* Mechanisms in Biological Competition. Sympos. Soc. Exp. Biol. 15, Cambridge Univ. Press, Cambridge, U.K.

74. Buse, L.J.; Baker, W.D. 1991. Determining necessity and priority for tending in young spruce plantations in northwestern Ontario. Ont. Min. Nat. Resour., Northwestern Ont. For. Technol. Devel. Unit, Thunder Bay ON, Tech. Note TN-08. 4 p.

75. Simard, S. 1966. 1996. Mixtures of paper birch and conifers: An ecological balancing act. pp. 15-22 In: P.G. Comeau & K.D. Thomas (eds.) Silviculture of temperate boreal broadleaf-conifer mixtures. BC Ministry of Forests, Research Branch, Victoria, BC. Land Management Handbook 36.

76. Green, D.S. 2004. Describing condition-specific determinants of competition in boreal and sub-boreal mixedwood stands. For. Chron. 80(6):736–742.

77. Lieffers, V.J.; Pinno, B.; Stadt, K.J. 2002. Light dynamics and free-to-grow standards in aspen-dominated mixedwood forests. For. Chron. 78(1):137–145.

78. Bella, I.E. 1971. A new competition model for individual trees. For. Sci. 17:364–372.

79. Hegyi, F. 1974. A simulation model for managing jack pine stands. p. 74–90 *in* Fries, J. (Ed.). Growth Models for Tree and Stand Simulation. R. Coll. For., Dep. For. Yield Res., Res. Note 30.

80. Ek, A.R.; Monserud, R.A. 1974. Trials with program FOREST: Growth and reproduction simulation of mixed species forest stands. p. 56–73 *in* Fries, J. (Ed.). Growth Models for Tree and Stand Simulation. R. Coll. For., Dep. For.

Yield Res., Res. Note 30.

81. Howard, K.M.; Newton, M. 1984. Overtopping by successional coast range vegetation slows Douglas-fir seedlings. J. For. 82:178–180.

82. Daniels, R.F. 1976. Simple competition indices and their correlation with annual loblolly pine tree growth. For. Sci. 22:454–456.

83. Wagner, R.G. 1982. A method for assessing severity of weeds in young plantations. *in* The Use of Herbicides in Forest Vegetation Management. Co-ordinated Research on Alternative Forestry Treatments and Systems (CRAFTS). Univ. Oregon, Dep. For. Sci., Portland OR. (Cited by Morris and MacDonald 1991, orig. not seen).

84. Weiner, J. 1984. Neighbourhood interference amongst *Pinus rigida* individuals. J. Ecol. 72:183–195.

85. Zedaker, S.M. 1982. Growth and development of young Douglas-fir in relation to intra- and inter-specific competition. Oregon State Univ., Corvallis OR, Ph.D. thesis. (Cited by Morris and MacDonald 1991, orig. not seen.)

86. Brand, D.G.; Kehoe, P.; Connors, M. 1986. Coniferous afforestation leads to soil acidification in central Ontario. Can. J. For. Res. 16(6):1389–1391.

87. Willcocks, A.[J.] and Bell, W. 1995. How initial forest plantation density affects future stand growth. OMNR, Northeast Science & Technology, NEST Tech. Note TN-008. 16 p.

88. Smith, D.M. 1962. The Practice of Silviculture, 7th ed. Wiley & Sons, New York NY. 378 p.

89. [OMNR] Ontario Mnistry of Natural Resources. 1989. Operational guidelines for tree improvement in Ontario. Ont. Min. Nat. Resour., Toronto ON, 9 sections separately paginated.

90. Drew, J.T.; Flewelling, J.W. 1979. Stand density management: an alternative approach and its application to Douglas-fir plantations. For. Sci. 25:518–532.

91. Archibald, D.J.; Bowling, C. 1995. Jack pine density management diagram for boreal Ontario. Ont. Min. Nat. Resour., Northeast Science & Technology, Tech. Note TN-005 NWST TN-31. 20 p.

92. Smith, N.J.; Brand, D.G. 1988. Compatible growth models and stand density diagrams. p. 636–643 *in* Ek. A.R..; Shifley, S.R..; Burk, T.E. (Eds.). Forest Growth Modelling and Prediction. Vol. 2. Proc. IUFRO Conf., August 1987, Minneapolis MN.

93. *Kittredge, J (1929). "Forest planting in the Lake States". U.S.D.A., For. Serv., Washington DC, Agric. Bull. (1497): 87 p.*

94. 5 Keefe,K. 2012;

95. 'Fact Sheet 4.12. Forest Regeneration', *IPCC Special Report on Land Use, Land-Use Change And Forestry*

96. Daniel, Theodore, John Helms, and Frederick Baker. *Principles of Silviculture*. 2nd ed. New York: McGraw-Hill, 1979. Print.

97. Lloyd, G.D. 1991. Juvenile spacing in rgw sub-boreal spruce biogeoclimatic region of British Columbia. p. 20–26 *in* Haavisto, V.F.; Smith, C.R.; Mason, C. (Eds.). Space to grow: spacing and thinning in northern Ontario. Proc. sympos., June 1990, Sault Ste. Marie ON, For. Can., Ont. Region/Ont. Min. Nat. Resour. Joint Rep. 15.

98. Hermelin, J. 1991. Spacing in eastern Canada: the New Brunswick experience. p. 27–29 *in* Haavisto, V.F., Smith, C.R.; Mason, C. (Eds.). Space to grow: spacing and thinning in northern Ontario. Proc. Sympos., June 1990, Sault Ste. Marie ON, For. Can., Ont. Region/Ont. Min. Nat. Resour. Joint Rep. 15.

99. Nicks. B.D. 1991. A company perspective on spacing and thinning. p. 62–67 *in* Haavisto, V.F.; Smith, C.R.; Mason, C. (Eds.). Space to grow: spacing and thinning in northern Ontario. Proc. sympos., June 1990, Sault Ste. Marie ON, For. Can., Ont. Region/Ont. Min. Nat. Resour. Joint Rep. 15.

100. MacKay, T. 1991. Reducing stem density with herbicides: what are the options?. p. 99–103 *in* Haavisto, V.F.; Smith, C.R.; Mason, C. (Eds.). Space to grow: spacing and thinning in northern Ontario. Proc. sympos., June 1990, Sault Ste. Marie ON, For. Can., Ont. Region/Ont. Min. Nat. Resour. Joint Rep. 15.

101. Ker, M.F. 1981. Early response of balsam fir to spacing in northwestern New Brunswick. Environment Canada, Canadian Forestry Service, Maritimes Forestry Centre, Fredericton, New Brunswick, Information Report M-X-129, 36 p.

102. Bella, I.E. 1986. Spacing effects 20 years after planting three conifers in Manitoba. Can., Can. For. Serv., Edmonton AB, For. Manage. Note No. 39. 11 p.

103. Stiell, W.M.; Berry, A.B. 1973. Development of unthinned white spruce plantations to age 50 at Petawawa Forest Experiment Station. Can. Dep. Environ., Can. For. Serv., Ottawa ON, Publ. 1317. 18 p..

104. Pollack, J.C.; Johnstone, W.; Coates, K.D.; LePage, P. 1992. The influence of initial espacement on the growth of a 32-year-old white spruce plantation. B.C. Min. For., Victoria BC, Res. Note 111. 16 p.

105. Bella, I.E.; De Franceschi, J.P. 1980. Spacing effects 15 years after planting three conifers in Manitoba. Can. Dep. Environ., Can. For. Serv., Northern For. Res. Centre, Edmonton AB, Inf. Rep. NOR-X-223. 10 p.

106. 6 Savill,P.S. 2004;

107. Navratil, S.; Branter, K.; and Zasada, J. 1991. Regeneration in the mixedwoods. p. 32–48 *in* Shortreid, A. (Ed.), Proc. Northern Mixedwood ‘89: a symposium at Fort St. John, B.C., Sept. 1989. For. Can., Pacific For. Centre, Victoria BC, FRDA Report 164.

108. 7 D’Amato,Anthony W. 2011;

109. Day, M.W. 1967. Pre-commercial thinning in conifers with silvicides. Michigan State Univ., Agric. Exp. Sta., East Lansing MI, Quarterly Bull 50:59–62.

110. The Dictionary of Forestry, The Society of American Foresters

111. Smith, D.M., B.C. Larson, M.J. Kelty, and P.M.S. Ashton. *The Practice of Silviculture: Applied Forest Ecology.* 9th edition. New York: John Wiley &

Sons, Inc., 1997. Print.

112. Haddock, P.G. 1961. Silvicultural views on the Canadian spruce forests. For. Chron. 37(4):376–389.

113. Wright, T.G. 1959. The Canadian spruce forest. For. Chron. 35(4):291–297.

114. Davis, L.S.; Johnson, K.N. 1987. Forest Management, 3rd ed. McGraw-Hill, New York, NY. 790 p.

115. Burgess, D.; Larocque, G.R.; Brand, D.G. 2001. Forest growth and future yields: the importance of today's regeneration practices. P. 603–624 *in* Wagner, R.G.; Colombo, S.J. (Eds.). Regenerating the Canadian Forest: Principles and Practice for Ontario. Fitzhenry & Whiteside, Markham ON in co-operation with Ont. Min. Nat. Resour. 650 p.

116. Hearnden, K.W.; Millson, S.V.; Wilson, W.C. 1996. Ontario Independent Forest Audit Committee. Ont. Min. Nat. Resour., Sault Ste. Marie ON. 117 p.

117. Waldron, R.M. 1966. Factors affecting natural white spruce regeneration on prepared seedbeds at the Riding Mountain Forest Experimental Area, Manitoba. Can. Dep. For. Rural Devel., For. Branch, Ottawa ON, Deptl. Publ. 1169. 41 p.

118. Butt, G.; Bancroft, B.; Folk, R. 1989. Ingress of Engelmann spruce and subalpine fir in southern interior ESSF. For. Can./B.C. Min. For., Victoria BC, Project 3.61, unpub. rep. (Cited in Coates et al. 1994)

119. Youngblood, A.P.; Zasada, J.C. 1991. White spruce artificial regeneration options on river floodplains in interior Alaska. Can. J. For. Res. 21(4):423–433.

120. Day, M.W.; Rudolph, V.J. 1970. Development of a white spruce plantation. Michigan State Univ., Agric. Exp. Sta., East Lansing MI, Res. Pap. 111. 4 p.

121. Herring, L.J.; McMinn, R.G. 1980. Natural and advanced regeneration of Engelmann spruce and subalpine fir compared 21 years after site treatment. For. Chron. 56:55–57.

122. McMinn, R.G. 1986. Comparative productivity of seedbed, natural and planted regeneration following various site treatments in white spruce clearcuts. p. 31–33 *in* Murray, M. (Ed.), The Yield Advantages of Artificial Regeneration at High Latitudes. Proc. 6th Internat. Workshop on Forest Regeneration. USDA, For. Serv., Pacific Northwest For. Range Exp. Sta., Gen. Tech. Rep. PNW-194. 60 p. (Cited in Coates et al. 1994).

123. Roussel, L. 1948. Couvert et photométrie. Bull. Soc. Forest Franche-Comté 25:313–326. For. Abs. 10:458–459, 1949.

124. Reifsnyder, W.E.; Lull, H.W. 1965. Radiant Energy in Relation to Forests. USDA, For. Serv. Washington DC, Tech. Bull. 1344. 111 p.

125. Oregon State University Extension Service

126. RESIDUAL DAMAGE IN A CONIFER STAND THINNED WITH A CTL SYSTEM, University of Idaho

127. 8 Bauhus,Jürgen 2009;

128. Sauder, E.A. 1996. Techniques to maintain a windfirm understory. p. 31 *in* FERIC West, Vancouver BC, Work Program 1996.

129. *Effects of Group-Selection Opening Size on Breeding Bird Habitat Use in a Bottomland Forest.* Christopher E. Moorman and David C. Guynn Jr. Ecological Applications , Vol. 11, No. 6 (Dec., 2001), pp. 1680-1691. Web. 4 Oct. 2013.

130. 9 Schulte,Benedict J. 1998;

131. 12 Gamborg,Christian 2003;

132. Kubin, E.; Kemppainen, L. 1991. Effect of clearcutting of boreal spruce forest on air and soil temperature conditions. Acta Forestalia Fennica No. 225. 42 p.

133. Smith, D.M., B.C. Larson, M.J. Kelty, P.M.S. Ashton (1997) The Practice of Silviculture: Applied Forest Ecology, John Wiley & Sons, p. 340-46

134. 10 Harmer,R. 2004;

135. Prochnau, A.E. 1963. Direct seeding experiments with white spruce, alpine fir, Douglas fir and lodgepole pine in the central interior of British Columbia. B.C. Min. For. Lands, Victoria BC, MOFL No. 37, Res Note. 24 p. (Cited in Coates et al. 1994).

136. Smith, J.H.G.; Clark, M.B. 1960. Growth and survival of Engelmann spruce and alpine fir on seed spots at Bolean Lake, B.V. 1954–59. For. Chron. 36(1):46–49, 51. (Cited in Coates et al. 1994).

137. Biolley, H. 1920. L'aménagement des forêts par la méthode expérimentale et spécialement la méthode du contrôle. Paris, Attinger frères. 90 p.

138. Helliwell, R., and E. R. Wilson. (2012). Continuous cover forestry in Britain: challenges and opportunities..Quarterly Journal of Forestry 106(3): 214-224

139. Kittredge, J. Jr.; Gevorkiantz, S. R. 1929. Forest possibilities of aspen lands in the Lake States. Minnesota Agricultural Exp. Sta., Minneapolis, Minnestota, Technical Bulletin 60.

140. Gardner, A.C. 1980. Regeneration problems and options for white spruce on river floodplains in the Yukon Territory. p. 19–24 *in* Murray, M.; Van Veldhuizen, R.M. (Eds.). Forest Regeneration at High,

141. Weetman, G.; Vyse, A. 1990. Natural regeneration. p. 118–130 *in* Lavender, D.P.; Parish, R.; Johnson, C.M.; Montgomery, G.; Vyse, A.; Willis, R.A.; Winston, D. (Eds.). Regenerating British Columbia's Forests. Univ. B.C. Press, Vancouver BC. (Cited in Coates et al. 1994).

142. Glew, D.R. 1963. The results of stand treatment in the white spruce–alpine fir type of the northern interior of British Columbia. B.C. Dep. Lands For. Water Resour., B.C. For. Serv., Victoria BC, For. Manage. Note 1. 27 p. (Cited in Coates et al. 1994).

143. Zasada, J.C. 1972. Guidelines for obtaining natural regeneration of white spruce in Alaska. USDA, For. Serv., Pacific Northwest For. Range Exp. Sta., Portland OR. 16 p.

144. Lähde, F. and Tuohisaari, O. 1976. An ecological study on the effects of shelters on germination and germling development of Scotch pine, Norway spruce and

Siberian larch. Comm. Inst. For. Reprint 88.1. 35 p.

145. Putman, W.E.; Zasada, J.C. 1986. Direct seeding techniques to regenerate white spruce in interior Alaska. Can. J. For. Res. 16(3):660–664.

146. Putman, W.E.; Zasada, J.C. 1985. Raven damage to plastic seeding shelters in interior Alaska. North. J. Appl. For. 2(2):41–43. (Cited in Coates et al. 1994).

147. Dyck, J.R. 1994. Converting aspen stands to white spruce–aspen mixedwoods by planting and seeding, Manitoba: Manitoba forestry demonstration areas. Can./ Manitoba Partnership in For., unumbered rep. 28 p.

148. 3 Brose,Patrick H. 2008; 3 Brose,Patrick H. 2008;

149. 4 Holgén,Per 2000; 4 Holgén,Per 2000;

150. *Survival and Growth of Under-Planted Trees: A Meta-Analysis across Four Biomes*. Alain Paquette, André Bouchard and Alain Cogliastro Ecological Applications , Vol. 16, No. 4 (Aug., 2006), pp. 1575-1589

151. 4 Holgén,Per 2000;

152. Baldwin, V.C. 1977. Regeneration following shelterwood cutting in a New Brunswick softwood stand. Can. Dep. Fish. & Environ., Can. For. Serv., Fredericton NB, Inf. Rep. M-X-76. 23 p.

153. Alexander, R.R. 1973. Partial cutting in old-growth spruce-fir. USDA For. Serv., Rocky Mountain and Range Exp. Sta., Fort Collins, CO, Research Paper RM-100. 16 p.

154. *Conservation Approaches for Woody, Early Successional Communities in the Eastern United States*. Frank R. Thompson, III and Richard M. DeGraaf Wildlife Society Bulletin , Vol. 29, No. 2 (Summer, 2001), pp. 483-494. Web. 4 October 2013.

155. Robertson, W.M. 1927. Cutting for reproduction in spruce stands. For. Chron 3(3):7–10.

156. Scott, J.D. 1970. Direct seeding in Ontario. For. Chron. 46(6):453–457.

157. Butt, G. 1988. Backlog forest land rehabilitation in the SBS and BWBS zones in the northern interior of British Columbia. Can./B.C. FRDA Rep. 023. 125 p.

158. Dyer, E.D.A.; Taylor, D.W. 1968. Attractiveness of logs containing female spruce beetles *Dendroctontus obesus* (ColeopteraL Scolytidae). Can. Entomol 100: 769-776.

159. DeLong, C. 1991. Dynamics of boreal mixedwood ecosystems. p.30–31 *in* Northern Mixedwood ‘89: Proceedings of a symposium held at Fort St. John, B.C., Sept. 1989. A. Shortreid (Ed.), For. Can., Pacific For. Centre, Victoria BC, FRDA Report 164.

160. Kohm, K. A, and Franklin, J. F., Creating a forestry for the 21st century: the science of ecosystem management. Island Press. 1997, ISBN 978-1-55963-399-4

161. *The Role of Disturbance in the Ecology and Conservation of Birds* Jeffrey D. Brawn, Scott K. Robinson and Frank R. Thompson III Annual Review of Ecology and Systematics , Vol. 32, (2001), pp. 251-276. Web. 4 October 2013.

162. *Effects of Selection Cutting on Bird Communities in Contiguous Eastern Hardwood Forests* . Andrew P. Jobes, Erica Nol and Dennis R. Voigt The Journal of Wildlife Management , Vol. 68, No. 1 (Jan., 2004), pp. 51-60. Web. 4 October 2013.

163. Brace, L.; Bella, I. 1988. Understanding the understorey: dilemma and opportunity. *In* White spruce understories. Canada–Alberta Agreement, Projects 1480, 1488, 20204. For. Can., Northern For. Centre, Edmonton AB. (Cited in Coates et al. 1994)

164. Sauder, E.A. 1990. Mixedwood harvesting. section B In White spruce understories, Canada–Alberta Agreement, Projects 1480, 1488, 20204. For. Can., Northern For. Centre, Edmonton AB, various pagination.

165. Sauder, E.A.; Sinclair, A.W.J. 1989. Harvesting in the mixedwood forest. Paper included in White spruce understories, Canada–Alberta Agreement, Projects 1480, 1488, 20204. For. Can., Northern For. Centre, Edmonton AB.

166. Brace, L. 1989. Protecting understorey white spruce when harvesting aspen. *In* White spruce understories, Canada–Alberta Agreement, Projects 1480, 1488, 20204. For. Can., Northern For. Centre, Edmonton AB.

167. Brace, L. 1990. A test of three logging systems in Alberta. Paper included in White spruce understories, Canada–Alberta Agreement, Projects 1480, 1488, 20204. For. Can., Northern For. Centre, Edmonton AB.

168. Stewart, J.D.; Landhäusser, S.M.; Stadt, K.J.; Lieffers, V.J. 2001. Predicting natural regeneration of white spruce in boreal mixedwood understories. For. Chron. 77(6):1006–1013.

169. McCaughey, W.W.; Schmidt, W.C. 1982. Understory tree release following harvest cutting in spruce–fir forests of the Intermountain West. USDA, For. Serv., Intermountain For. Range Exp. Sta., Res. Pap. INT-285. Cited in Coates et al. 1994).

170. Johnstone, W.D. 1978. Growth of fir and spruce advance growth and logging residuals following logging in west-central Alberta. Can. Dep. Environ., Can. For. Serv., Northern For. Res. Centre, Edmonton AB, Inf. Rep. NOR-X-203. 16 p.

171. Ball, W.J.; Kolabinski, V.S. 1979. An aerial reconnaissance of softwood regeneration on mixedwood sites in Saskatchewan. Can. Dep. Environ., Can. For. Serv., Northern For. Res. Centre, Edmonton AB, Inf. Rep. NOR-X-216. 14 p.

172. McCaughey, W.W.; Ferguson, D.E. 1988. Response of advance regeneration to release in the Inland Mountain West: a summary. p. 255–266 *in* Schmidt, W.C. (Compiler). Proc. Future Forests of the Mountain West: A Stand Culture Symp. , Sept./Oct. 1986, Missoula MT. USDA, For. Serv., Intermount. Res. Sta., Ogden UT, Gen. Tech. Rep. INT-243. 402 p.

173. Smith, R.B.; Wass, E.F. 1979. Tree growth on and adjacent to contour skid roads in the subalpine zone, southeastern British Columbia. Can. Dep. Environ., Can. For. Serv., Victoria BC, Report BC-R-2. 26 p.

174. Stettler, R.F. 1958. Development of a residual stand of interior spruce–alpine fir during the first twenty-eight years following cutting to a 12-inch diameter limit. For. Serv., Victoria BC, Res. Note 34. 15 p. [Coates et al. 1994]

175. Baskerville, G.L. 1961. Response of young fir and spruce to release from shrub competition. Can. Dep. Northern Affairs and National Resources, For. Branch, Ottawa ON, For. Res. Div., Tech. Note 98. 14 p. (Cited in Coates et al. 1994)

176. McKinnon, L.M.; Mitchell, A.K.; Vyse, A. 2002. The effects of soil temperature and site preparation on subalpine and boreal tree species: a bibliography. Nat. Resour., Can., Can. For. Serv., Victoria BC, Inf. Rep. BC-X-394. 29 p.

177. Macadam, A.M. 1987. Effects of broadcast slash burning on fuels and soil chemical properties in the sub-boreal spruce zone of central British Columbia. Can. J. For. Res. 17(12):1577–1584.

178. Kiil, A.D.; Chrosciewicz, Z. 1970. Prescribed fire – its place in reforestation. For. Chron. 46:448–451.

179. Holt, L. 1955. White spruce seedbeds as related to natural regeneration. Pulp Paper Res. Instit. Can., Montreal QC. 28 p.

180. Ballard, T.M. 1985. Spruce nutrition problems in the central interior and their relationship with site preparation. Proc. Interior spruce seedling performance: state of the art Symposium. Northern Silviculture Committee Workshop, Feb. 1985, Prince George BC.

181. Tarrant, R.F. 1954. Effect of slash burning on soil pH. USDA, For. Serv., Pacific Northwest For. and Range Exp. Sta., Portland OR, Res. Note 102. 5 p.

182. Taylor, S.W.; Feller, M.C. 1987. Initial effects of slashburning on the nutrient status of Sub-boreal Spruce Zone ecosystems. *In* Papers presented at the Fire Management Symposium, April 1987, Prince George BC, Central Interior Fire Protection Committee, Smithers BC.

183. Little, S.N.; Klock, G.O. 1985. The influence of residue removal and prescribed fire on distribution of forest nutrients. USDA, For. Serv., Res. Pap. PNW-333.

184. Endean, F.; Johnstone, W.D. 1974. Prescribed fire and regeneration on clearcut spruce–fir sites in the foothills of Alberta. Environ. Can., Can. For. Serv., Northern For. Res. Centre, Edmonton AB, Inf. Rep. NOR-X-126. 33 p.

185. Kiil, A.D. 1965. Weight and size distribution of slash of white spruce and lodgepole pine. For. Chron. 41:432–437.

186. Kiil, A.D. 1968. Weight of the fuel complex in 70-year-old lodgepole pine stands of different densities. Department of Forestry and Rural Development, Forest Research Laboratory, Calgary, Alberta. Departmental Publication 1228. 13 p.

187. Ronco, F. 1975. Diagnosis: sunburned trees. J. For. 73(1):31–35. (Cited in Coates et al. 1994).

188. Wang, G.G.; Siemens, A.; Keenan, V.; Philippot, D. 2000. Survival and growth of black and white spruce seedlings in relation to stock type, site preparation and plantation type in southeastern Manitoba. For. Chron. 76(5):775–782.

189. Hall, J. 1970. Site preparation in Ontario. For. Chron. 46:445–447.

190. Draper, D.; Binder, W.; Fahlman, R.; Spittlehouse, D. 1985. Post-planting ecophysiology of Interior spruce. Interior Spruce Seedling Performance: State of the Art. Northern Silvic. Committee, Prince George BC. 18 p. (mimeo).

191. Burton, P.; Bedford, L.; Goldstein, M.; Osberg, M. 2000. Effect of disk trench orientation and planting spot position on the ten-year performance of lodgepole pine. New For. 20:23–44.

192. Clausen, J.C.; Mace, A.C., Jr. 1972. Accumulation and snowmelt on north–south versus east–west oriented clearcut strips. Univ. Minnesota, Coll. For., St. Paul MN, Minn. For. Res. Notes No. 34. 4 p.

CHAPTER 7

Wildlife Management

Chapter Contents

- **Introduction**
- **History**
- **Wildlife Conservation**
- **Species Reintroduction**

The term wildlife means different things to different people. To a backyard wildlifer, it may mean chickadees, nuthatches, and cardinals. To a hunter, it may mean white-tailed deer, bobwhite quail, and gray squirrels. To a sheep producer, it may mean coyotes. To a poultry producer, it may mean mink, weasels, skunks, and raccoons. To a gardener, it may mean hummingbirds and butterflies.

INTRODUCTION

Wildlife management attempts to balance the needs of wildlife with the needs of people using the best available science. Wildlife management can include game keeping, wildlife conservation and pest control. Wildlife management draws on disciplines such as mathematics, chemistry, biology, ecology, climatology and geography to gain the best results.

Wildlife conservation aims to halt the loss in the Earth's biodiversity by taking into consideration ecological principles such as carrying capacity, disturbance and succession and environmental conditions such as physical geography, pedology and hydrology with the aim of balancing the needs of wildlife with the needs of people. Most wildlife biologists are concerned with the preservation and improvement of habitats although rewilding is increasingly being used. Techniques can include reforestation, pest control, nitrification and denitrification, irrigation, coppicing and hedge laying.

Game keeping is the management or control of wildlife for the well being of game and may include killing other animals which share the same niche or predators to maintain a high population of the more profitable species, such as pheasants introduced into woodland. In his 1933 book *Game Management*, Aldo Leopold, one of the pioneers of wildlife management as a science, defined it as "the art of making land produce sustained annual crops of wild game for recreational use".

Pest control is the control of real or perceived pests and can be used for the benefit of wildlife, farmers, game keepers or safety reasons. In the United States, wildlife management practices are often implemented by a governmental agency to uphold a law, such as the Endangered Species Act.

In the United Kingdom, wildlife management undertaken by several organizations including government bodies such as the Forestry Commission, Charities such as the RSPB and The Wildlife Trusts and privately hired gamekeepers and contractors. Legislation has also been passed to protect wildlife such as the Wildlife and Countryside Act 1981. The UK government also give farmers subsidies through the Countryside Stewardship Scheme to improve the conservation value of their farms.

Wildlife traditionally refers to undomesticated animal species, but has come to include all plants, fungi, and other organisms that grow or live wild in an area without being introduced by humans.

Wildlife can be found in all ecosystems. Deserts, forests, rain forests, plains, grasslands, and other areas including the most developed urban areas, all have distinct forms of wildlife. While the term in popular culture usually refers to animals that are untouched by human factors, most scientists agree that much wildlife is affected by human activities.

Humans have historically tended to separate civilization from wildlife in a number of ways including the legal, social, and moral sense. Some animals, however, have adapted to suburban environments. This includes such animals as domesticated cats, dogs, mice, and gerbils. Some religions declare certain animals to be sacred, and in modern times concern for the natural environment has provoked activists to protest against the exploitation of wildlife for human benefit or entertainment.

The global wildlife population has decreased by 52 percent between 1970 and 2014, according to a report by the World Wildlife Fund.

HISTORY

Game Laws

The history of wildlife management begins with the game laws, which regulated the right to kill certain kinds of fish and wild animal (game). In Britain game laws developed out of the forest laws, which in the time of the Norman kings were very oppressive. Under William the Conqueror, it was as great a crime to kill one of the king's deer as to kill one of his subjects. A certain rank and standing, or the possession of a certain amount of property, were for a long time qualifications indispensably necessary to confer upon any one the right of pursuing and killing game.

The Game Act of 1831 protected game birds by establishing close seasons when they could not be legally taken. The act made it lawful to take game only with the provision of a game licence and provided for the appointment of gamekeepers around the country. The purposes of the law was to balance the needs for preservation and harvest and to manage both environment and populations of fish and game.

The Game Act 1831 protects game birds in England and Wales.

Early game laws were also enacted in the US; - in 1839 Rhode Island closed the hunting season for white-tailed deer from May to November. Other regulations during this time focused primarily on restricting hunting. At this time, lawmakers did not consider population sizes or the need for preservation or restoration of wildlife habitats.

Emergence of Wildlife Conservation

The late 19th century saw the passage of the first pieces of wildlife conservation legislation and the establishment of the first nature conservation societies. The Sea Birds Preservation Act of 1869 was passed in Britain as the first nature protection law in the world after extensive lobbying from the Association for the Protection of Seabirds.

The Royal Society for the Protection of Birds was founded as the Plumage League in 1889 by Emily Williamson at her house in Manchester as a protest group campaigning against the use of great crested grebe and kittiwake skins and feathers in fur clothing. The group gained popularity and eventually amalgamated with the Fur and Feather League in Croydon to form the RSPB. The Society attracted growing support from the suburban middle-classes as well as support from many other influential figures, such as the ornithologist Professor Alfred Newton.

The National Trust formed in 1895 with the manifesto to "...promote the permanent preservation, for the benefit of the nation, of lands, ...to preserve (so far practicable) their natural aspect." On 1 May 1899, the Trust purchased two acres of Wicken Fen with a donation from the amateur naturalist Charles Rothschild, establishing the first nature reserve in Britain. Rothschild was a pioneer of wildlife conservation in Britain, and went on to establish many other nature reserves, such as one at Woodwalton Fen, near Huntingdon, in 1910. During his lifetime he built and managed his estate at Ashton Wold in Northamptonshire to maximise its suitability for wildlife, especially butterflies. Concerned about the loss of wildlife habitats, in 1912 he set up the Society for the Promotion of Nature Reserves, the forerunner of The Wildlife Trusts partnership.

During the society's early years, membership tended to be made up of specialist naturalists and its growth was comparatively slow. The first independent Trust was formed in Norfolk in 1926 as the Norfolk Naturalists Trust, followed in 1938 by the Pembrokeshire Bird Protection Society which after several subsequent changes of name is now the Wildlife Trust of South and West Wales and it was not until the 1940s and 1950s that more Naturalists' Trusts were formed in Yorkshire, Lincolnshire, Leicestershire and Cambridgeshire. These early Trusts tended to focus on purchasing land to establish nature reserves in the geographical areas they served.

Wildlife Management in the US

The profession of wildlife management was established in the United States in the 1920s and '30s by Aldo Leopold and others who sought to transcend the purely restrictive policies of the previous generation of conservationists, such as anti-hunting activist William T. Hornaday. Leopold and his close associate Herbert Stoddard, who had both been trained in scientific forestry, argued that modern science and technology could be used to restore and improve wildlife habitat and thus produce abundant "crops" of ducks, deer, and other valued wild animals.

The institutional foundations of the profession of wildlife management were established in the 1930s, when Leopold was granted the first university professorship in wildlife management (1933, University of Wisconsin, Madison), when Leopold's textbook 'Game Management' was published (1933), when The Wildlife Society was founded, when the Journal of Wildlife Management began publishing, and when the first Cooperative Wildlife Research Units were established. Conservationists planned many projects throughout the 1940s. Some of which included the harvesting of female mammals such as deer to decrease rising populations. Others included waterfowl and wetland research. The Fish and Wildlife Management Act was put in place to urge farmers to plant food for wildlife and to provide cover for them.

In 1937, the Federal Aid in Wildlife Restoration Act (also known as the Pittman-Robertson Act) was passed in the U.S.. This law was an important advancement in the field of wildlife management. It placed a 10% tax on sales of guns and ammunition. The funds generated were then distributed to the states for use in wildlife management activities and research. This law is still in effect today.

Wildlife management grew after World War II with the help of the GI Bill and a postwar boom in recreational hunting. An important step in wildlife management in the United States national parks occurred after several years of public controversy regarding the forced reduction of the elk population in Yellowstone National Park. In 1963, United States Secretary of the Interior Stewart Udall appointed an advisory board to collect scientific data to inform future wildlife management. In a paper known as the Leopold Report, the committee observed that culling programs at other national parks had been ineffective, and recommended active management of Yellowstone's elk population.

Elk overpopulation in Yellowstone is thought by many wildlife biologists, such as Douglas Smith, to have been primarily caused by the extirpation of wolves from the park and surrounding environs. After wolves were removed, elk herds increased in population, reaching new highs during the mid-1930s. The increased number of elk apparently resulted in overgrazing in parts of Yellowstone. Park officials decided that the elk herd should be managed. For approximately thirty years, the park elk herds were culled: Each year some were captured and shipped to other locations, a certain number were killed by park rangers, and hunters were allowed to take more elk that migrated outside the park. By the late 1960s the herd populations dropped to historic lows (less than 4,000 for the Northern Range herd). This caused outrage among both conservationists and hunters. The park service stopped culling elk in 1968. The elk population then rebounded. Twenty years later there were 19,000 elk in the Northern Range herd, a historic high.

Since the tumultuous 1970s, when animal rights activists and environmentalists began to challenge some aspects of wildlife management, the profession has been overshadowed by the rise of conservation biology. Although wildlife managers remain central to the implementation of the Endangered Species Act and other wildlife conservation policies, conservation biologists have shifted the focus of conservation away from wildlife management's concern with the protection and restoration of single species and toward the maintenance of ecosystems and biodiversity.

Types of wildlife management

There are n a population, either changing its numbers by direct means or influencing numbers by the indirect means of altering food supply, habitat, density of predators, or prevalence of disease. This is appropriate when a population is to be harvested, or when it slides to an unacceptably low density or increases to an unacceptably high level. Such densities are inevitably the subjective view of the land owner, and may be disputed by animal welfare interests.

- **Custodial management** is preventive or protective. The aim is to minimize external influences on the population and its habitat. It is appropriate in a national park where one of the stated goals is to protect ecological processes. It is also appropriate for conservation of a threatened species where the threat is of external origin rather than being intrinsic to the system. Feeding of animals by visitors is generally discouraged.

Opposition

The control of wildlife through killing and hunting has been criticized by animal rights and animal welfare activists. Critics object to the real or perceived cruelty involved in some forms of wildlife management.

Environmentalists have also opposed hunting where they believe it is unnecessary or will negatively affect biodiversity.Critics of game keeping note that habitat manipulation and predator control are often used to maintain artificially inflated populations of valuable game animals (including introduced exotics) without regard to the ecological integrity of the habitat.

Game keepers in the UK claim it to be necessary for wildlife conservation as the amount of countryside they look after exceeds by a factor of nine the amount in nature reserves and national parks.

Management of Hunting Seasons

Wildlife management studies, research and lobbying by interest groups help designate times of the year when certain wildlife species can be legally hunted, allowing for surplus animals to be removed. In the United States, hunting season and bag limits are determined by guidelines set by the United States Fish and Wildlife Servicefor migratory game such as waterfowl and other migratory gamebirds. The hunting season and bag limits for state regulated game species such as deer are usually determined by State game Commissions, which are made up of representatives from various interest groups, wildlife biologists, and researchers.

Open and closed season on deer in the UK is legislated for in the Deer Act 1991 and the Deer Act (Scotland) 1996

Open Season

Open season is when wildlife is allowed to be hunted by law and is usually not during the breeding season. Hunters may be restricted by sex, age or class of animal, for instance there may be an open season for any male deer with 4 points or better on at least one side.

Limited Entry

Where the number of animals taken is to be tightly controlled, managers may have a type of lottery system called limited. Many apply, few are chosen. These hunts may still have age, sex or class restrictions.

Closed Season

Closed season is when wildlife is protected from hunting and is usually during its breeding season. Closed season is enforced by law, any hunting during closed season is punishable by law and termed as illegal hunting or poaching.

Type of Weapon Used

In wildlife management one of the conservation principles is that the weapon used for hunting should be the one that causes the least damage to the animal and is sufficiently effective so that it hits the target. Given State and Local laws, types of weapon can also vary depending on type, size, sex of game and also the geographical layout of that specific hunting area.

Food, Pets, and Traditional Medicines

A mesh bag full of live frogs waiting for a buyer at Chiang Mai's Thanin market. Frog meat in Thailand is mostly used in stir-fries and Thai curries.

Anthropologists believe that the Stone Age people and hunter-gatherers relied on wildlife, both plants and animals, for their food. In fact, some species may have been hunted to extinction by early human hunters. Today, hunting, fishing, and gathering wildlife is still a significant food source in some parts of the world. In other areas, hunting and non-commercial fishing are mainly seen as a sport or recreation, with the edible meat as mostly a side benefit of it. Meat sourced from wildlife that is not traditionally regarded as game is known as bush meat. The increasing demand for wildlife as a source of traditional food in East Asia is decimating populations of sharks, primates, pangolins and other animals, which they believe have aphrodisiac properties.

In November 2008, almost 900 plucked and "oven-ready" owls and other protected wildlife species were confiscated by the Department of Wildlife and National Parks in Malaysia, according to TRAFFIC. The animals were believed to be bound for China, to be sold in wild meat restaurants. Most are listed in CITES (the Convention on International Trade in Endangered Species of Wild Fauna and Flora) which prohibits or restricts such trade.

A November 2008 report from biologist and author Sally Kneidel, PhD, documented numerous wildlife species for sale in informal markets along the Amazon River, including wild-caught marmosets sold for as little as $1.60 (5 Peruvian soles).Many Amazon species, including peccaries, agoutis, turtles, turtle eggs, anacondas, armadillos, etc., are sold primarily as food. Others in these informal markets, such as monkeys and parrots, are destined for the pet trade, often smuggled into the United States. Still other Amazon species are popular ingredients in traditional medicines sold in local markets. The medicinal value of animal parts is based largely on superstition.

Religion

Many animal species have spiritual significance in different cultures around the world, and they and their products may be used as sacred objects in religious rituals. For example, eagles, hawks and their feathers have great cultural and spiritual value to Native Americans as religious objects. In Hinduism the cow is regarded sacred.

Muslims conduct sacrifices on Eid-ul-Adha to commemorate the sacrificial spirit of Ibrahim [Abraham] in love of God. Camels, sheep, goats, and cows may be offered as sacrifice during the three days of Eid.

Tourism

Many nations have established their tourism sector around their natural wildlife. South Africa has, for example, many opportunities for tourists to see the country's wildlife in its national parks, such as the Kruger Park. In South India the Periar Wildlife Sanctuary, Bandipur National Park and Mudamalai Wildlife Sanctuary are situated around and in forests. India is home to many national parks and wildlife sanctuaries showing the diversity of its wildlife, much of its unique fauna, and excels in the range. There are 89 national parks, 13 bio reserves and more than 400 wildlife sanctuaries across India which are the best places to go to see tigers, lions, elephants, rhinoceros, birds, and other wildlife which reflect the importance that the country places on nature and wildlife conservation.

Destruction

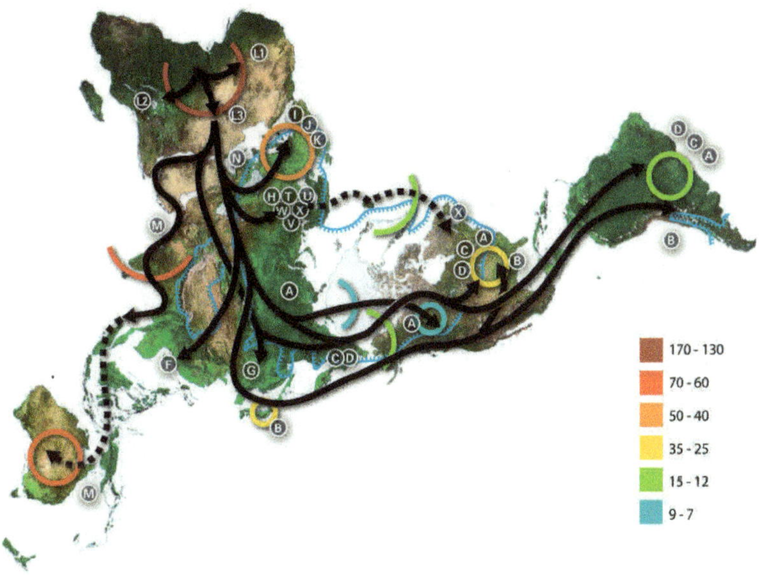

Map of early human migrations, according to mitochondrialpopulation genetics. Numbers are millennia before the present.

This subsection focuses on anthropogenic forms of wildlife destruction.

Exploitation of wild populations has been a characteristic of modern man since our exodus from Africa 130,000 – 70,000 years ago. The rate of extinctions of entire species of plants and animals across the planet has been so high in the last few hundred years it is widely believed that we are in the sixth great extinction event on this planet; the Holocene Mass Extinction.

Destruction of wildlife does not always lead to an extinction of the species in question, however, the dramatic loss of entire species across Earth dominates any review of wildlife destruction as extinction is the level of damage to a wild population from which there is no return.

The four most general reasons that lead to destruction of wildlife include overkill, habitat destruction and fragmentation, impact of introduced species and chains of extinction.

Overkill

Wildlife is an invaluable treasure but it is being exploited due to illegal trade of many of its species.Overkill happens whenever hunting occurs at rates greater than the reproductive capacity of the population is being exploited. The effects of this are often noticed much more dramatically in slow growing populations such as many larger species of fish. Initially when a portion of a wild population is hunted, an increased availability of resources (food, etc.) is experienced increasing growth and reproduction as density dependent inhibition is lowered. Hunting, fishing and so on, has lowered the competition between members of a population. However, if this hunting continues at rate greater than the rate at which new members of the population can reach breeding age and produce more young, the population will begin to decrease in numbers.

Populations that are confined to islands, whether literal islands or just areas of habitat that are effectively an "island" for the species concerned, have also been observed to be at greater risk of dramatic population declines following unsustainable hunting.

Habitat Destruction and Fragmentation

Deforestation and increased road-building in the Amazon Rainforest are a significant concern because of increased human encroachment upon wild areas, increased resource extraction and further threats to biodiversity.

The habitat of any given species is considered its preferred area or territory. Many processes associated with human habitation of an area cause loss of this area and decrease the carrying capacity of the land for that species. In many cases these changes in land use cause a patchy break-up of the wild landscape. Agricultural land frequently displays this type of extremely fragmented, or relictual, habitat. Farms sprawl across the landscape with patches of uncleared woodland or forest dotted in-between occasional paddocks.

Examples of habitat destruction include grazing of bushland by farmed animals, changes to natural fire regimes, forest clearing for timber production and wetland draining for city expansion.

Impact of Introduced Species

Mice, cats, rabbits, dandelions and poison ivy are all examples of species that have become invasive threats to wild species in various parts of the world[citation needed]. Frequently species that are uncommon in their home range become out-of-control invasions in distant but similar climates. The reasons for this have not always been clear and Charles Darwin felt it was unlikely that exotic species would ever be able to grow abundantly in a place in which they had not evolved. The reality is that the vast majority of species exposed to a new habitat do not reproduce successfully. Occasionally, however, some populations do take hold and after a period of acclimation can increase in numbers significantly, having destructive effects on many elements of the native environment of which they have become part.

Chains of Extinction

This final group is one of secondary effects. All wild populations of living things have many complex intertwining links with other living things around them. Large herbivorous animals such as the hippopotamus have populations of insectivorous birds that feed off the many parasitic insects that grow on the hippo. Should the hippo die out, so too will these groups of birds, leading to further destruction as other species dependent on the birds are affected. Also referred to as a domino effect, this series of chain reactions is by far the most destructive process that can occur in any ecological community.

Another example is the black drongos and the cattle egrets found in India. These birds feed on insects on the back of cattle, which helps to keep them disease-free. Destroying the nesting habitats of these birds would cause a decrease in the cattle population because of the spread of insect-borne diseases.

Media

The Douglas Squirrel.

Wildlife has long been a common subject for educational television shows. National Geographic specials appeared on CBS beginning in 1965, later moving to ABC and then PBS. In 1963, NBC debuted *Wild Kingdom,* a popular program featuring zoologist Marlin Perkins as host. The BBC natural history unit in the UK was a similar pioneer, the first wildlife series LOOK presented by Sir Peter Scott, was a studio-based show, with filmed inserts. It was in this series that David Attenborough first made his appearance which led to the series Zoo Quest during which he and cameraman Charles Lagus went to many exotic places looking for and filming elusive wildlife—notably the Komodo dragon in Indonesia and lemurs in Madagascar. Since 1984, the Discovery Channel and its spin off Animal Planet in the US have dominated the market for shows about wildlife on cable television, while on PBS the NATURE strand made by WNET-13 in New York and NOVA by WGBH in Boston are notable. See also Nature documentary. Wildlife television is now a multimillion-dollar industry with specialist documentary film-makers in many countries including UK, US, New Zealand NHNZ, Australia, Austria, Germany, Japan, and Canada. There are many magazines which cover wildlife including National Wildlife Magazine, Birds & Blooms, Birding (magazine), and Ranger Rick (for children).

WILDLIFE CONSERVATION

The Siberian tiger is a subspecies of tiger that is endangered; three subspecies of tiger are already extinct.

Conservation is the practice of protecting wild plant and animal species and their habitats. The goal of wildlife conservation is to ensure that nature will be around for future generations to enjoy and also to recognize the importance of wildlife and wilderness for humans and other species alike. Many nations have government agencies and NGO's dedicated to wildlife conservation, which help to implement policies designed to protect wildlife. Numerous independent non-profit organizations also promote various wildlife conservation causes.

According to the National Wildlife Federation, wildlife in the United States gets a majority of their funding through appropriations from the federal budget, annual federal and state grants, and financial efforts from programs such as the Conservation Reserve Program, Wetlands Reserve Program and Wildlife Habitat Incentives Program. Furthermore, a substantial amount of funding comes from the state through the sale of hunting/fishing licenses, game tags, stamps, and excise taxes from the purchase of hunting equipment and ammunition, which collects around $200 million annually.

Wildlife conservation has become an increasingly important practice due to the negative effects of human activity on wildlife. An endangered species is defined as a population of a living species that is in the danger of becoming extinct because of several reasons. Some of The reasons can be, that 1. the species have a very low population, or 2. they are threatened by the varying environmental or prepositional parameters.

Major Dangers to Wildlife

Fewer natural wildlife habitat areas remain each year. Moreover, the habitat that remains has often been degraded to bear little resemblance to the wild areas which existed in the past.Habitat loss—due to destruction, fragmentation and degradation of habitat—is the primary threat to the survival of wildlife in the United States. When an ecosystem has an ecosystem are some of the ways habitats can become so degraded that they no longer support native wildlife.

- Climate change: Global warming is making hot days hotter, rainfall and flooding heavier, hurricanes stronger and droughts more severe. This intensification of weather and climate extremes will be the most visible impact of global warming in our everyday lives. It is also causing dangerous changes to the landscape of our world, adding stress to wildlife species and their habitat. Since many types of plants and animals have specific habitat requirements, climate change could cause disastrous loss of wildlife species. A slight drop or rise in average rainfall will translate into large seasonal changes. Hibernating mammals, reptiles, amphibians and insects are harmed and disturbed. Plants and wildlife are sensitive to moisture change so, they will be harmed by any change in moisture level. Natural phenomena like floods, earthquakes, volcanoes, lightning, forest fires.
- Unregulated Hunting and poaching: Unregulated hunting and poaching causes a major threat to wildlife. Along with this, mismanagement of forest department and forest guards triggers this problem.
- Pollution: Pollutants released into the environment are ingested by a wide variety of organisms. Pesticides and toxic chemical being widely used, making the environment toxic to certain plants, insects, and rodents.
- Perhaps the largest threat is the extreme growing indifference of the public to wildlife, conservation and environmental issues in general. Over-exploitation

of resources, i.e., exploitation of wild populations for food has resulted in population crashes (over-fishing and over-grazing for example).

- Over exploitation is the over use of wildlife and plant species by people for food, clothing, pets, medicine, sport and many other purposes. People have always depended on wildlife and plants for food, clothing, medicine, shelter and many other needs. But today we are taking more than the natural world can supply. The danger is that if we take too many individuals of a species from their natural environment, the species may no longer be able to survive. The loss of one species can affect many other species in an ecosystem. The hunting, trapping, collecting and fishing of wildlife at unsustainable levels is not something new. The passenger pigeon was hunted to extinction, early in the last century, and over-hunting nearly caused the extinction of the American bison and several species of whales.
- Deforestation: Humans are continually expanding and developing, leading to an invasion of wildlife habitats. As humans continue to grow they clear forested land to crewe more space. This stresses wildlife populations as there are fewer homes and food sources to survive off of.

Population: The increasing population of human beings is the most major threat to wildlife. More people on the globe means more consumption of food,water and fuel . Therefore,more waste is generated. Every major threat to wildlife as seen above, is directly related to increasing population of human beings. If the population is altered so is the amount of risk to wildlife. The less is the population, less is the disturbance to wildlife.

Today, the [Endangered Species Act] protects some U.S. species that were in danger from over exploitation, and the Convention on International Trade in Endangered Species of Fauna and Flora (CITES) works to prevent the global trade of wildlife. But there are many species that are not protected from being illegally traded or over-harvested.

Wildlife conservation as a government involvement

In 1972, the Government of India enacted a law called the Wild Life (Protection) Act.The World Conservation Strategy was developed in 1980 by the "International Union for Conservation of Nature and Natural Resources" (IUCN) with advice, cooperation and financial assistance of the United Nations Environment Programme (UNEP) and the World Wildlife Fund and in collaboration with the Food and Agriculture Organization of the United Nations (FAO) and the United Nations Educational, Scientific and Cultural Organization (Unesco)" The strategy aims to "provide an intellectual framework and practical guidance for conservation actions." This thorough guidebook covers everything from the intended "users" of the strategy to its very priorities. It even includes a map section containing areas that have large seafood consumption and are therefore endangered by over fishing. The main sections are as follows:

The marking off of a sea turtle nest. Anna Maria, FL. 2012.

- The objectives of conservation and requirements for their achievement:
    - Maintenance of essential ecological processes and life-support systems.
    - Preservation of genetic diversity that is flora and fauna.
    - Sustainable utilization of species and ecosystems.
- Priorities for national action:
    - A framework for national and sub-national conservation strategies.
    - Policy making and the integration of conservation and development.
    - Environmental planning and rational use allocation.
- Priorities for international action:

- International action: law and assistance.
- Tropical forests and dry lands.
- A global programme for the protection of genetic resource areas.

Map sections:

- Tropical forests
- Deserts and areas subject to desertification.

Non-Government Involvement

As major development agencies became discouraged with the public sector of environmental conservation in the late 1980s, these agencies began to lean their support towards the "private sector" or non-government organizations (NGOs). In a World Bank Discussion Paper it is made apparent that "the explosive emergence of nongovernmental organizations" was widely known to government policy makers. Seeing this rise in NGO support, the U.S. Congress made amendments to the Foreign Assistance Act in 1979 and 1986 "earmarking U.S. Agency for International Development (USAID) funds for biodiversity". From 1990 moving through recent years environmental conservation in the NGO sector has become increasingly more focused on the political and economic impact of USAID given towards the "Environment and Natural Resources". After the terror attacks on the World Trade Centers on September 11, 2001 and the start of former President Bush's War on Terror, maintaining and improving the quality of the environment and natural resources became a "priority" to "prevent international tensions" according to the Legislation on Foreign Relations Through 2002 and section 117 of the 1961 Foreign Assistance Act. Furthermore, in 2002 U.S. Congress modified the section on endangered species of the previously amended Foreign Assistance Act.

SPECIES REINTRODUCTION

Species reintroduction is the deliberate release of a species into the wild, from captivity or other areas where the animal survives. A species that needs reintroduction is usually one whose existence has become threatened or endangered in the wild. However, reintroduction of a species can also be for pest control. For example, wolves being reintroduced to a wild area because of an overpopulation of elk or deer. Because reintroduction may involve returning native species to localities where they had been extirpated, some prefer the term "**reestablishment**".

Reintroduction has been practiced for many years.

Principles of Reintroduction

Reintroduction and translocation are both important tools for population and species management. Translocation moves wild-caught animals from one natural location to another, while reintroduction moves captive-born animals into their natural historical range. "Much of the behavioral research required to obtain reproduction will also be critical for reintroduction, which depends on the development of behaviorally competent individuals. More behavioral research to select the best candidates and prepare them for the very different challenges that await them in nature will be essential also. Behaviorists will again play a crucial role in post-release monitoring to determine the behavioral deficiencies that limit the success of reintroductions. There may be no other conservation action where the skills of behavioral researchers are more essential than reintroduction." In situ conservation means "on site". In-situ conservation is the conservation of species diversity within normal and natural habitats and ecosystems. The challenge in using in-situ methods is to expand our vision of protected areas to include multiple use and extractive reserves to develop new models for conservation. In-Situ conservation uses innovative proposals such as damaged ecosystems to preserve rare, endangered, and threatened species and to expand the range of options available for economic development. Ex-situ conservation means literally, "off-site conservation". It is the process of protecting an endangered species of plant or animal outside of its natural habitat. Capturing and relocating part of a population from a threatened habitat and placing it in a new location where it may have a better chance of survival is one example. Ex-situ conservation should only be used when In-Situ conservation cannot.

Education

Zoos are one of the most conventional methods of ex-situ conservation. Depending on their size and location, zoos receive between a few tens of thousands to several million visitors annually. Zoos provide education to the public about the many endangered species and explain the factors contributing the threats they face in their native habitats. Through ex-situ conservation methods they provide protected specimens for breeding and reintroduction into the wild. It has been suggested that this method should be used only when necessary and when In-Situ conservation is not possible.

Successes and Failures

Reintroduction biology is new and continues to be a work in progress. Although there is still no general and broadly accepted definition of reintroduction success, it has been proposed that the criteria widely used to assess the conservation status of endangered taxa, such as the IUCN Red List criteria, should be used to asses reintroduction success. In other words, successful reintroduction programs should yield viable and self-sustainable populations on the long-term. The IUCN/SSC Re-introduction Specialist Group & Environment Agency, in their 2011 Global Re-introduction Perspectives, compiled reintroduction case studies from around the world. 184 case studies were reported on a range of species which included invertebrates, fish, amphibians, reptiles, birds, mammals, and plants. Assessments from all of the studies included goals, success indicators, project summary, major difficulties faced, major lessons learned, and success of project with reasons for success or failure. The Siberian tiger population has rebounded from 40 individuals in the 1940s to around 500 in 2007. The Siberian tiger population is now the largest un-fragmented tiger population in the world. Yet, a high proportion of translocations and reintroductions have not been successful in establishing viable populations Black footed ferrets in North America have also been a highly successful reintroduced species. Husbandry techniques were refined and the first successful captive breeding occurred in 1987. For this species, disease management in captivity is paramount and strict protocols are followed. Genetic management is closely monitored and ferrets are bred using a mean-kinship strategy which provides genetic diversity when bred in captivity. To prepare them for reintroduction into their new "wild" habitat, captive-born ferrets are placed in outdoor pens, simulating quasi-natural conditions with dirt burrows and live prey. Ferrets that receive pre-release conditioning in outdoor pens prior to release have demonstrated 10-fold higher survival rates in the wild than ferrets that receive no pre-release exposure. Ferrets are typically released in the fall during their natural dispersal and kits (young of the year) are the primary candidates.

Many factors can attribute to the success or failure of a reintroduction. Predators, food, pathogens, competitors, and weather can all affect a reintroduced population's ability to grow, survive, and reproduce. Animals raised in captivity may experience stress during captivity or translocation, which can weaken their immune systems. The IUCN reintroduction guidelines emphasize the need for an assessment of the availability of suitable habitat as a key component of reintroduction planning.Poor assessment of the release site can increase the chances that the species will reject the site and perhaps move to a less suitable environment. This can decrease the species fitness and thus decrease chances for survival. They state that restoration of the original habitat and amelioration of causes of extinction must be explored and considered as essential conditions for these projects. Unfortunately, the monitoring period that should follow re introductions often remains neglected.

Improving Research Techniques

A cooperative approach to reintroduction by ecologists and biologists could improve research techniques. For both preparation and monitoring of reintroductions, increasing contacts between academic population biologists and wildlife managers is encouraged within the Survival Species Commission and the IUCN. The IUCN states that a re-introduction requires a multidisciplinary approach involving a team of persons drawn from a variety of backgrounds. A survey by Wolf et al. in 1998 indicated that 64% of reintroduction projects have used subjective opinion to assess habitat quality. This means that most reintroduction evaluation has been based on human anecdotal evidence and not enough has been based on statistical findings. Seddon et al. (2007) suggest that researchers contemplating future reintroductions should specify goals, overall ecological purpose, and inherent technical and biological limitations of a given reintroduction, and planning and evaluation processes should incorporate both experimental and modeling approaches.

Monitoring the health of individuals, as well as the survival, is important; both before and after the reintroduction. Intervention may be necessary if the situation proves unfavorable. Population dynamics models that integrate demographic parameters and behavioral data recorded in the field can lead to simulations and tests of a priori hypotheses. Using previous results to design further decisions and experiments is a central concept of adaptive management. In other words, learning by doing can help in future projects. Population ecologists should therefore collaborate with biologists, ecologists, and wildlife management to improve reintroduction programs.

Survival Skills

It may be very hard to reintroduce species into the wild, even if their natural habitats were restored. Survival techniques, which are normally passed from parents to offspring during parenting, are lost. The genetics of the species is saved, but the natural memetics of the species is not.

Beginning in the 1980s, biologists have learned that many mammals and birds need to learn a lot to survive in the wild. Thus, reintroduction programmes have to be planned carefully, ensuring that the animals have the necessary survival skills. Biologists must also study the animals after the reintroduction to learn whether the animals are surviving and breeding, what effects the reintroduction has on the ecosystem, and how to improve the process.

Still, a vast number of animals may need to be reintroduced into the wild to be sure that enough of them learn how to survive. For instance, in reintroducing houbara bustards into the wild in the United Arab Emirates, more than 5,000 birds per year are used.

Re-introduction Specialist Group (RSG)

The RSG is a network of specialists whose aim is to combat the ongoing and massive loss of biodiversity by using re-introductions as a responsible tool for the management and restoration of biodiversity. It does this by actively developing and promoting sound inter-disciplinary scientific information, policy, and practice to establish viable wild populations in their natural habitats. The role of the RSG is to promote the re-establishment of viable populations in the wild of animals and plants. The need for this role was felt due to the increased demand from re-introduction practitioners, the global conservation community and increase in re-introduction projects worldwide.

Increasing numbers of animal and plant species are becoming rare, or even extinct in the wild. In an attempt to re-establish populations, species can – in some instances – be re-introduced into an area, either through translocation from existing wild populations, or by re-introducing captive-bred animals or artificially propagated plants.

United Kingdom

Ongoing or Successful Programs

Reserva CIBE.

- European bison to the UK – (ongoing)
- Black grouse to Derbyshire – (ongoing)
- Common crane to Somerset – (ongoing)
- Corncrake to Cambridgeshire – (ongoing)
- Glanville fritillary butterfly to Somerset – (successful)
- European beaver to Scotland, England, and Wales – (successful in Scotland and Northern England)
- Great bustard to Salisbury Plain – (ongoing)
- Heath fritillary butterfly to Essex – (successful)
- Ladybird spider to Arne RSPB reserve in Dorset, England – (ongoing).
- Large blue butterfly in the West and The South West – (successful and ongoing)
- Northern goshawk – the existing UK population is believed to be derived from a mixture of escaped falconers' birds and deliberate introductions – (successful)
- Osprey to Rutland Water – (successful)
- Red kite in the Chiltern Hills, Black Isle, Northamptonshire, Dumfries and Galloway, Yorkshire, Perth and Kinross and Gateshead – (successful)
- Red squirrel to Anglesey – (successful and ongoing)
- Reindeer to the Cairngorms in Scotland – (semi-domesticated; successful)
- Scots pine to southern England – (unplanned, successful)
- Silver-washed fritillary to Essex – (ongoing, locally successful)
- Wild boar to several places in Britain – (accidental, successful)
- White-tailed eagle to the Hebrides – (successful)
- White-tailed eagle to the east coast of Scotland – (ongoing)

Planned or Proposed Programs

- Brown bear in the Scottish Highlands – (proposed)
- Elk to Great Britain – (proposed)
- European lynx in Wales, England and Scotland (proposed)
- Golden eagle in United States and Wales
- Gray wolf in Scotland – (proposed)
- Arctic wolf in northern Scotland
- White stork – (proposed)
- White-tailed eagle to England and Wales (planned – on hold while a suitable site is found)
- Gray whale to Atlantic Ocean and Irish Sea (proposed – considered not feasible technologically and biologically. Two vagrant individuals sighted in the Mediterranean Sea in 2010 and off Namibia in 2013, the latter is the first confirmation of the species in the Southern Hemisphere.)

Other Countries

Planned or Proposed Programmes

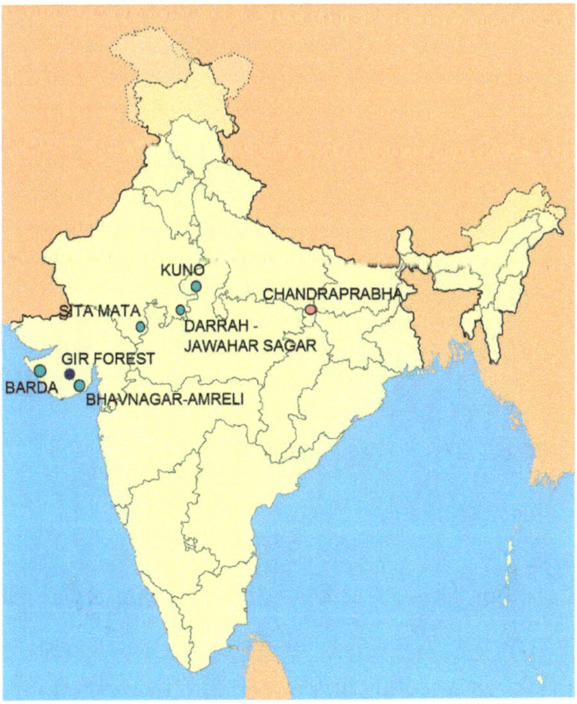

Lion reintroduction sites in India.

- Asiatic Lion Reintroduction Project of Asiatic lion to Kuno Wildlife Sanctuary from their only home presently in the world at Gir Forest National Park. Kuno Wildlife Sanctuary is the chosen site for re-introducing and establishing the world's second completely separate population of the wild free ranging Asiatic lions in the state of Madhya Pradesh. It was decided to re-introduce the Asiatic lion in Kumbhalgarh Wildlife Sanctuary in Rajasthan. Some will be reintroduced in two locations in Gujarat.
- Siberian Tiger Re-population Project was proposed in 2009 to reintroduce Amur tigers back to their former lands and including the former ranges in Central Asia once inhabited by their closest relatives, the Caspian tigers. In 2010, two pairs of Siberian tigers, exchanged for Persian leopards to southwestern Russia, were set to be reintroduced in Iran's Miankaleh peninsula. Currently, the big cats (one of them had died) are being held in captivity in Eram zoo. Siberian tigers were also proposed to be reintroduced to a suitable habitat near the international river of Amu Darya in Central Asia and near the Ili River delta in Kazakhstan. A rewilding project at the Pleistocene Park, part of the re-population project was proposed back in 2005.
- Red deer - A programme was announced in 2013 to reintroduce the red deer to Armenia. 4 males and 11 females of the species will be purchased and transported to a breeding centre at Dilijan National Park. The World Wildlife Fund Germany and Orange Armenia have provided the funds for the project.
- Cheetah reintroduction in India is a project to reintroduce the cheetah in India. The Asiatic cheetah became extinct in 1947 when Maharaja of Surguja hunted the last three in the state of Rewa in central India. It was officially declared extinct in 1952 by the Indian government. Plans are going on to reintroduce the cheetah to two site in Madhya Pradesh (Kuno Wildlife Sanctuary and Nauradehi Wildlife Sanctuary) and in Rajasthan's Shahgarh Landscape. However, the Supreme court of India put a hold on this project as they recommended to protect the endangered lions first rather than import cheetah from Africa or Iran. (On hold)
- South China tiger - Captive tigers being re-wilded in Laohu Valley Reserve in the Free State province of South Africa under Save China's Tigers programme, will be eventually released back into the wilderness of China.

Ongoing or Successful Programs

Africa

- North African ostrich in Morocco, Nigeria, Niger and Tunisia (ongoing)
- Southern white rhinoceros in Kenya, Uganda and Zambia (successful)
- South African cheetah in Swaziland (successful)

Asia

- Amur leopard in Russia (successful)
- Arabian oryx in the Sultanate of Oman (successful), United Arab Emirates (successful), Israel (successful)
- Bornean orangutan in East Kalimantan, Indonesia
- Indian rhinoceros in Pakistan (ongoing)
- North African ostrich in Israel and Saudi Arabia (ongoing)
- Nubian ibex in Israel (successful)
- Eurasian otter in Japan (ongoing)
- Père David's deer in China (successful)
- Persian fallow deer in Israel (ongoing)
- Persian leopard in Russia (ongoing)
- Persian onager in Saudi Arabia (successful)
- Przewalski's horse in Mongolia (ongoing)
- Sarus Cranes in Thailand (ongoing)
- Sudan cheetah in United Arab Emirates (ongoing)
- Turkmenian kulan in Kazakhstan and Uzbekistan (successful)
- Yarkon Bleak fish in Israel (successful)
- Neurergus microspilotus in [Western Iran] (successful)

Europe

- Alpine ibex in the French, Italian and Swiss Alps (successful)
- Eurasian brown bear in the Alps (ongoing)
- European beaver in several places in Europe (successful)
- European otter in the Netherlands (ongoing)
- European lynx in Switzerland (successful), and other parts of Europe (ongoing)
- European black vulture in the Massif Central in France
- Goitered gazelle in Protected Areas of Vashlovani in Georgia (country)(ongoing)
- Golden eagle in Ireland (ongoing)
- Griffon vulture in the Massif Central, France (successful), Central Apennines, Italy, and Northern and Southern Israel (ongoing)
- Iberian lynx in Portugal (ongoing)
- Lammergeier in the Alps (successful) Switzerland (successful)
- Lesser kestrel in Spain
- Lesser white-fronted goose in Sweden and Germany (ongoing)

- Northern bald ibis in Austria and Italy (ongoing)
- Peregrine falcon in Germany, Poland, Sweden and Norway
- Red kite in Ireland
- Western swamphen in the Mondego River basin, Portugal (successful)
- White-tailed eagle in Ireland (ongoing)
- Wisent in Poland, Belarus (successful) and other parts of Europe (ongoing)

North America

- Black-footed ferret in Canada, United States and Mexico
- Blanding's Turtle in Canada
- California condor in California and Mexico (ongoing)
- Fisher in Washington State (ongoing)
- Grey wolf to Yellowstone National Park (successful)
- Musk ox in Alaska (United States) (successful)
- American flamingo to Anegada, British Virgin Islands (successful)
- Whooping Cranes, including migratory population in the Eastern United States and non-migratory population in Louisiana (ongoing)

Oceania

- Woylie in Australia (ongoing)

South America

- Giant anteater in Corrientes, Argentina
- Guanaco in central Argentina
- Puerto Rican parrot in Arecibo (ongoing)
- Red-and-green macaw to Iberá Provincial Reserve, Argentina
- Vinaceous amazon parrot to Parque Nacional das Araucárias, Santa Catarina, Brazil (ongoing)

REFERENCES

1. Potter, Dale R.; Kathryn M. Sharpe; John C. Hendee (1973). Human Behavior Aspects Of Fish And Wildlife Conservation - An Annotated Bibliography (PDF). U.S. Dept. of Agriculture. p. 290.
2. M. E. Soulé and B. A. Wilcox. 1980. Conservation Biology: An Evolutionary-Ecological Perspective. Sinauer Associatess. Sunderland, Massachusetts.
3. M. E. Soule. (1986). What is conservation Biology? BioScience, 35(11): 727-734

4. Soule, Michael E. (1986). Conservation Biology: The Science of Scarcity and Diversity. Sinauer Associates. p. 584. ISBN 9780878937950.
5. Hunter, M. L. (1996). Fundamentals of Conservation Biology. Blackwell Science Inc., Cambridge, Massachusetts., ISBN 0-86542-371-7.
6. Groom, M.J., Meffe, G.K. and Carroll, C.R. (2006) Principles of Conservation Biology (3rd ed.). Sinauer Associates, Sunderland, MA. ISBN 0-87893-518-5
7. van Dyke, Fred (2008). Conservation Biology: Foundations, Concepts, Applications, 2nd ed. Springer Verlag. p. 478. ISBN 978-1-4020-6890-4.
8. Wildlife Conservation Management (PDF).
9. Bolen, Eric G., Robinson, William. (1999). Wildlife Ecology and Management. Prentice Hall.
10. G. Baeyens, M. L. Martinez (2007). Coastal Dunes: Ecology and Conservation. Springer. p. 282.
11. Protecting seabirds at Bempton Cliffs".
12. Milestones. RSPB. Retrieved 19 February 2007.
13. History of the RSPB. RSPB. Retrieved 19 February 2007.
14. Wicken Fen National Nature Reserve. Wicken Fen..
15. Wildlife Trust for Bedfordshire, Cambridgeshire, Northamptonshire and Peterborough
16. Pevsner, Nikolaus (1961). The Buildings of England – Northamptonshire. London and New Haven: Yale University Press. pp. 94–5. ISBN 978-0-300-09632-3.
17. Leopold, A. Starker, et al. 1963. "The Goal of Park Management in the United States. Wildlife Management in the National Parks. National Park Service. Retrieved on September 19, 2009.
18. League Against Cruel Sports - Consequences of the Shoot
19. RSPB - Which birds are threatened
20. National Gamekeepers' Organisation Charitable Trust

CHAPTER 8

Forest Restoration

Chapter Contents

- **Introduction**
- **Scope**
- **Opportunities for Forest Restoration**
- **Natural Regeneration**
- **Post-Fire Regeneration**
- **Forest Restoration Projects**
- **Forest Landscape Restoratio**
- **Secondary Forest**
- **Restoration Ecology**

The need to repair habitat and restore forest structure and funciton is recognized throughout the temperate and boreal zones as a component of sustainable forest management. Forest restoration is a complex task, complicated by diverse ecological and social conditions, that challenges our understanding of forest ecosystems. The term restoration is used indiscriminately and it is difficult to define in a way that compasses all situations found in the literature and practive. Generall, restoration is seen as symmetric with degradation: an undisturbed forest in a natural or historical condition can be degraded, and a degraded forest can be restored to that natural or historical condition. As will become apparent, reality is more complicated and the fully restored state is probably unattainable. Terminology, however, is not merely an academic issue; definitions related to forestry and restoration are used under several international conventions such as climate change and biodiversity where distinctions and nuance have important policy implications .

INTRODUCTION

Forest restoration is defined as "actions to re-instate ecological processes, which accelerate recovery of forest structure, ecological functioning and biodiversity levels towards those typical of climax forest" i.e. the end-stage of natural forest succession. Climax forests are relatively stable ecosystems that have developed the maximum biomass, structural complexity and species diversity that are possible *within the limits imposed by climate and soil and without continued disturbance from humans* (more explanation here). Climax forest is therefore the target ecosystem, which defines the ultimate aim of forest restoration. Since climate is a major factor that determines climax forest composition, global climate change may result in changing restoration aims.

In the 1980s, conservation organizations warned that, once destroyed, tropical forests could never be restored. Thirty years of restoration research now challenge this: a) This site in Doi Suthep-Pui National Park, N. Thailand was deforested, over-cultivated and then burnt. The black tree stump was one of the original forest trees. Local people teamed up with scientists to repair their watershed.

b) Fire prevention, nurturing natural regeneration and planting framework tree species resulted in trees growing above the weed canopy within a year.

c) After 12 years, the restored forest overwhelmed the black tree stump.

Forest restoration is a specialized form of reforestation, but it differs from conventional tree plantations in that its primary goals are biodiversity recovery and environmental protection.

SCOPE

Forest restoration may include simply protecting remnant vegetation (fire prevention, cattle exclusion etc.) or more active interventions to accelerate natural regeneration, as well as tree planting and/or sowing seeds (direct seeding) of species characteristic of the target ecosystem. Tree species planted (or encouraged to establish) are those that are typical of, or provide a critical ecological function in, the target ecosystem. However, wherever people live in or near restoration sites, restoration projects often include economic species amongst the planted trees, to yield subsistence or cash-generating products.

Forest restoration is an inclusive process, which depends on collaboration among a wide range of stakeholders including local communities, government officials, non-government organizations, scientists and funding agencies. Its ecological success is measured in terms of increased biological diversity, biomass, primary productivity, soil organic matter and water-holding capacity, as well as the return of rare and keystone species, characteristic of the target ecosystem. Economic indices of success include the value of forest products and ecological services generated (e.g. watershed protection, carbon storage etc.), which ultimately contribute towards poverty reduction. Payments for such ecological services (PES) and forest products can provide strong incentives for local people to implement restoration projects.

OPPORTUNITIES FOR FOREST RESTORATION

Demonstration forest restoration plot, SUNY-ESF, Syracuse, NY.

Forest restoration is appropriate wherever biodiversity recovery is one of the main goals of reforestation, such as for wildlife conservation, environmental protection, ecotourism or to supply a wide variety of forest products to local communities. Forests can be restored in a wide range of circumstances, but degraded sites within protected areas are a high priority, especially where some climax forest remains as a seed source within the landscape. Even in protected areas, there are often large deforested sites: logged over areas or sites formerly cleared for agriculture. If protected areas are to act as Earth's last wildlife refuges, restoration of such areas will be needed.

Many restoration projects are now being implemented under the umbrella of "forest landscape restoration" (FLR), defined as a "planned process to regain ecological integrity and enhance human well-being in deforested or degraded landscapes". FLR recognizes that forest restoration has social and economic functions. It aims to achieve the best possible compromise between meeting both conservation goals and the needs of rural communities. As human pressure on landscapes increases, forest restoration will most commonly be practiced within a mosaic of other forms of forest management, to meet the economic needs of local people.

NATURAL REGENERATION

Tree planting is not always essential to restore forest ecosystems. A lot can be achieved by studying how forests regenerate naturally, identifying the factors that limit regeneration and devising methods to overcome them. These can include weeding and adding fertilizer around natural tree seedlings, preventing fire, removing cattle and so on. This is "accelerated" or "assisted" natural regeneration. It is simple and cost-effective, but it can only operate on trees that are already present, mostly light-loving pioneer species. Such tree species are not usually those that comprise climax forests, but they can foster recolonization of the site by shade-tolerant climax forest tree species, via natural seed dispersal from remnant forest. Because this is a slow process, biodiversity recovery can usually be accelerated by planting some climax forest tree species, especially large-seeded, poorly dispersed species. It is not feasible to plant all the tree species that may have formerly grown in the original primary forest and it is usually unnecessary to do so, if the framework species method can be used.

POST-FIRE REGENERATION

In large parts of the world, forest fires cover a heavy toll on forests. That can be because of provoked deforestation in order to substitute forests by crop areas, or in dry areas, because of wild fires occurring naturally or intentionally. A whole section of forest landscape restoration in linked to this particular problem, as in many cases, the net loss of ecosystem value is very high and can open the drop to an accelerated further degradation of the soil conditions through erosion and desertification. This indeed has dire consequences on both the quality of the habitats and their related fauna. Nevertheless, in some specific cases, wild fires do actually allow to increase the biodiversity index of the burnt area, in which case the Forest Restoration Strategies tend to look for a different land-use.

FOREST RESTORATION PROJECTS

Ashland Forest Resiliency Stewardship Project

The Ashland Forest Resiliency Stewardship Project (AFR) is a decade long, science-based project launched in 2010 with the intent of reducing severe wildfire risk, but also protecting water quality, old-growth forest, wildlife, people, property, and the overall quality of life within the Ashland watershed. The primary stakeholders in this cooperative restoration effort are the U.S. Forest Service, the City of Ashland, Lomaktsi Restoration Project, and the Nature Conservancy. The project was launched with initial funding from the Economic Recovery stimulus, and has more recently received funding from the Forest Service Hazardous Fuels program and the Joint Chiefs Landscape Restoration Partnerships program to back the project through 2016 .

Located in the dry forests of southern Oregon, the threat of wildfire is a reality for land managers and property owners alike. The boundaries of the city of Ashland intersect with the surrounding forest in what is referred to as the wildland-urban interface (WUI). Historically, the forests of this region experienced a relatively frequent fire return interval, which prevented buildup of heavy fuel loads. A century of fire exclusion and suppression on federal lands in the Pacific Northwest has led to increased forest density and fuel loads, and thus a more persistent threat of devastating wildfire.

The AFR project has implemented restoration techniques and prescriptions that aim to replicate the process of ecological succession in dry, mixed-conifer forests of the Pacific Northwest. The approach involves a combination of fuels reduction, thinning small-diameter trees, and carrying out prescribed burns. Priority is given to maintaining ecological function and complexity by retaining the largest and oldest trees, preserving wildlife habitat and riparian areas, and protecting erodible soils and maintaining slope stability.

Since its inception in 2010, the AFR project has completed restoration work on 4,000 of the 7,600 acres slated for the project. The project has provided educational experience to over 2,000 students and has benefitted the local community by creating jobs and providing workforce training. Currently, helicopter logging operations are thinning 1,100 acres of the watershed while controlled burning operations take place as air quality conditions allow.

FOREST LANDSCAPE RESTORATION

Forest landscape restoration (FLR) is defined as "a planned process to regain ecological integrity and enhance human well-being in deforested or degraded landscapes". It comprises tools and procedures to integrate site-level forest restoration actions with desirable landscape-level objectives, which are decided upon via various participatory mechanisms among stakeholders. The concept has grown out of collaboration among some of the world's major international conservation organizations including the International Union for Conservation of Nature (IUCN), the World Wide Fund for Nature (WWF), the World Resources Institute and the International Tropical Timber Organization (ITTO).

Aims

The concept of FLR was conceived to bring about compromises between meeting the needs of both humans and wildlife, by restoring a range of forest functions at the landscape level. It includes actions to strengthen the resilience and ecological integrity of landscapes and thereby keep future management options open. The participation of local communities is central to the concept, because they play a critical role in shaping the landscape and gain significant benefits from restored forest resources. Therefore, FLR activities are inclusive and participatory.

Desirable Outcomes

The desirable outcomes of an FLR program usually comprise a combination of the following, depending on local needs and aspirations:

- identification of the root causes of forest degradation and prevention of further deforestation,
- positive engagement of people in the planning of forest restoration, resolution of land-use conflicts and agreement on benefit-sharing systems,

- compromises over land-use trade-offs that are acceptable to the majority of stakeholders,
- a repository of biological diversity of both local and global value,
- delivery of a range of utilitarian benefits to local communities including:
    - a reliable supply of clean water,
    - environmental protection particularly watershed services (e.g. reduced soil erosion, lower landslide risk, flood/drought mitigation etc.),
    - a sustainable supply of a diverse range of forest products including foods, medicines, firewood etc.,
    - monetary income from various sources e.g. ecotourism, carbon trading via the REDD+ mechanism and from payments for other environmental services (PES)

Activities

FLR combines several existing principles and techniques of development, conservation and natural resource management, such as landscape character assessment, participatory rural appraisal, adaptive management etc. within a clear and consistent evaluation and learning framework. An FLR program may comprise various forestry practices on different sites within the landscape, depending on local environmental and socioeconomic factors. These may include protection and management of secondary and degraded primary forests, standard forest restoration techniques such as "assisted" or "accelerated" natural regeneration (ANR) and the planting of framework tree species to restore degraded areas, as well as conventional tree plantations and agroforestry systems to meet more immediate monetary needs

The IUCN hosts the The Global Partnership on Forest Landscape Restoration, which co-ordinates development of the concept around the world.

In 2014, the Food and Agricultural Organization of the United Nations established the Forest and Landscape Restoration Mechanism. The Mechanism supports countries to implement FLR as a contribution to achieving the Bonn Challenge - the restoration of 150 million hectare of deforested and degraded lands by 2020 - and the Convention on Biological Diversity Aichi Biodiversity Targets - related to ecosystem conservation and restoration.

In partnership with the Global Mechanism of the United Nations Convention to Combat Desertification, FAO released two discussion papers on sustainable financing for FLR in 2015. *Sustainable Financing for Forest and Landscape Restoration: The Role of Public Policy Makers* provides recommendations and examples of FLR financing for countrues. *Sustainable Financing for Forest and Landscape Restoration - Opportunities, challenges and the way forward* provides an overview of funding sources and financial instruments available for FLR activities.

SECONDARY FOREST

The forest in Stanley Park, Vancouver, Canada is generally considered to have second and third growth characteristics. This photo shows regeneration, a tree growing out of the stump of another tree that was felled in 1962 by the remnants of Typhoon Freda.

A **secondary forest** (or **second-growth forest**) is a forest or woodland area which has re-grown after a timber harvest, until a long enough period has passed so that the effects of the disturbance are no longer evident. It is distinguished from an old-growth forest (primary or primeval forest), which has not recently undergone such disruption, and complex early seral forest, as well as third-growth forests that result from harvest in second growth forests. Secondary forest regrowing after timber harvest differs from forest regrowing after natural disturbances such as fire, insect infestation, or windthrow because the dead trees remain to provide nutrients, structure, and water retention after natural disturbances. However, often after natural disturbance the timber is harvested and removed from the system, in which case the system more closely resembles secondary forest rather than complex early seral forest.

Description

Depending on the forest, the development of primary characteristics may take anywhere from a century to several millennia. Hardwood forests of the eastern United States, for example, can develop primary characteristics in one or two generations of trees, or 150–500 years. Often the disruption is the result of human activity, such as logging, but natural phenomena that produce the same effect are often included in the definition. Secondary forests tend to have trees closer spaced than primary forests and contain less undergrowth than primary forests. Secondary forests typically were thought to lack biodiversity compared to primary forests, however this has been challenged in recent years Usually, secondary forests have only one canopy layer, whereas primary forests have several.

Secondary forestation is common in areas where forests have been lost by the slash-and-burn method, a component of some shifting cultivation systems of agriculture. Secondary forests may also arise from forest that has been harvested heavily or over a long period of time, forest that is naturally regenerating from fire and from abandoned pastures or areas of agriculture. It takes a secondary forest typically forty to 100 years to begin to resemble the original old-growth forest; however, in some cases a secondary forest will not succeed, due to erosion or soil nutrient loss in certain tropical forests.

Secondary forests re-establish by the process of succession. Openings created in the forest canopy allow sunlight to reach the forest floor. An area that has been cleared will first be colonized by pioneer species. Even though some species loss may occur with primary forest removal, a secondary forest can protect the watershed from further erosion and provides habitat. Secondary forests may also buffer edge effects around mature forest fragments and increase connectivity between them. They may also be a source of wood and other forest products.

Today most of the forest of the United States, the eastern part of North America and Europe consist of secondary forest.

Rainforests

In the case of tropical rainforests, where soil nutrient levels are characteristically low, the soil quality may be significantly diminished following the removal of primary forest. In Panama, growth of new forests from abandoned farmland exceeded loss of primary rainforest in 1990. However, due to the diminished quality of soil, among other factors, the presence of a significant majority of primary forest species fail to recover in these second-growth forests.

RESTORATION ECOLOGY

Recently constructed wetland regeneration in Australia, on a site previously used for agriculture

Rehabilitation of a portion of Johnson Creek, to restore bioswale and flood control functions of the land which had long been converted to pasture for cow grazing. The horizontal logs can float, but are anchored by the posts. Just-planted trees will eventually stabilize the soil. The fallen trees with roots jutting into the stream are intended to enhance wildlife habitat. The meandering of the stream is enhanced here by a factor of about three times, perhaps to its original course.

Restoration ecology emerged as a separate field in ecology in the 1980s. It is the scientific study supporting the practice of **ecological restoration**, which is the practice of renewing and restoring degraded, damaged, or destroyed ecosystems and habitats in the environment by active human intervention and action. Restoration ecology is therefore commonly used for the academic study of the process, whereas ecological restoration is commonly used for the actual project or process by restoration practitioners.

The Society for Ecological Restoration defines *"ecological restoration"* as an "intentional activity that initiates or accelerates the recovery of an ecosystem with respect to its health, integrity and sustainability". The practice of ecological restoration includes wide scope of projects such as erosion control, reforestation, usage of genetically local native species, removal of non-native species and weeds, revegetation of disturbed areas, daylighting streams, reintroduction of native species, as well as habitat and range improvement for targeted species.

E. O. Wilson, a biologist states that: "Here is the means to end the great extinction spasm. The next century will, I believe, be the era of restoration in ecology.'

Restoration ecology is the scientific study of ecological restoration.

Land managers, laypeople, and stewards have been practicing ecological restoration or ecological management for many hundreds, if not thousands of years, yet the scientific field of "restoration ecology' was not first formally identified and coined until the late 1980s, by John Aber and William Jordan when they were at the University of Wisconsin-Madison. They held the first international meetings on this topic in Madison during which attendees visited the University of Wisconsin's Arboretum—the oldest restoration ecology project made famous by Professor Aldo Leopold. The study of restoration ecology has only become a robust and independent scientific discipline over the last two decades, and the commercial applications of ecological restoration have tremendously increased in recent years.

Restoration Needs

There is consensus in the scientific community that the current environmental degradation and destruction of many of the Earth's biota is considerable and is taking place on a "catastrophically short timescale". Estimates of the current extinction rate is 1000 to 10,000 times more than the normal rate. For many people biological diversity, (biodiversity) has an intrinsic value that humans have a responsibility towards other living things, and an obligation to future generations.

On a more anthropocentric level, natural ecosystems provide human society with food, fuel and timber. Fundamentally, ecosystem services involve the purification of air and water, detoxification and decomposition of wastes, regulation of climate, regeneration of soil fertility and pollination of crops. Such processes have been estimated to be worth trillions of dollars annually.

Habitat loss is the leading cause of both species extinctions and ecosystem service decline. The two ways to reverse this trend of habitat loss are conservation of currently viable habitat and restoration of degraded habitats.

Conservation biology and restoration ecology

Restoration ecology may be viewed as a sub-discipline of conservation biology, the scientific study of how to protect and restore biodiversity, and restoration a part of the resulting conservation movement.

Conservation vs. Restoration

The fundamental difference between restoration and other conservation efforts is analogous to the difference between disease prevention and treatment. Conservation attempts to maintain and protect existing habitat and biodiversity, whereas restoration attempts to reverse existing environmental degradation and population declines. Targeted human intervention is used to promote habitat, biodiversity recovery and associated gains.

The possibility of restoration, however, does not provide an excuse for converting extremely valuable "pristine' habitat into other uses: as in medicine, it better to prevent than to treat. "Treatment' is generally less effective and more expensive than prevention, and "treatment' cannot always restore the condition before the "injury': some habitat and biodiversity losses are permanent.

Focuses

Though restoration ecologists and other conservation biologists generally agree that habitat is the most important locus of biodiversity protection, the disciplines themselves have different focuses. Conservation biology as an academic discipline is rooted in population biology. Because of that, it is generally organized at the genetic level, looking at specific species populations (i.e. endangered species). Restoration ecology is organized at the community level, looking at specific ecosystems.

Because it is organized by species, conservation biology often emphasizes vertebrate animals because of their salience and popularity, whereas restoration ecology emphasizes plants because of their importance within ecosystems: ecosystem restoration is botanically based. Since soils define the foundation of any functional terrestrial system, restoration ecology's ecosystem-level focus also results in greater emphasis on the role of soil's physical and microbial processes.

Modes of Inquiry

Conservation biology's focus on rare or endangered species limit the number of manipulative studies that can be performed. As a consequence, conservation studies tend to be descriptive, comparative and unreplicable. However, the highly manipulative nature of restoration ecology allows the researcher to test the hypotheses vigorously. Restorative activity often reflects an experimental test of what limits populations.

Theoretical foundations

Restoration ecology draws on a wide range of ecological concepts.

Disturbance

Disturbance is a change of environmental conditions, which interferes with the functioning of a biological system. Disturbance, at a variety of spatial and temporal scales is a natural, and even essential, component of many communities.

Humans have had limited "natural' impacts on ecosystems for as long as humans have existed, however, the severity and scope of our influences has accelerated in the last few centuries. Understanding and minimizing the differences between modern anthropogenic and "natural' disturbances is crucial to restoration ecology. For example, new forestry techniques that better imitate historical disturbances are now being implemented.

In addition, restoring a fully sustainable ecosystem often involves studying and attempting to restore a natural disturbance regime (e.g., fire ecology).

Succession

Ecological succession is the process by which the component species of a community changes over time. Following a disturbance, an ecosystem generally progresses from a simple level of organization (i.e. few dominant species) to a more complex community (i.e. many interdependent species) over few generations. Depending on the severity of the disturbance, restoration often consists of initiating, assisting or accelerating ecological successional processes.

In many ecosystems, communities tend to recover following mild to moderate natural and anthropogenic disturbances. Restoration in these systems involves hastening natural successional trajectories. However, a system that has experienced a more severe disturbance (i.e. physical or chemical alteration of the environment) may require intensive restorative efforts to recreate environmental conditions that favor natural successional processes. This ability to recover is called resilience.

Fragmentation

Habitat fragmentation is the emergence of spatial discontinuities in a biological system. Through land use changes (e.g. agriculture) and "natural' disturbance, ecosystems are broken up into smaller parts. Small fragments of habitat can support only small populations and small populations are more vulnerable to extinction. Furthermore, fragmenting ecosystems decreases interior habitat. Habitat along the edge of a fragment has a different range of environmental conditions and therefore supports different species than the interior. Fragmentation effectively reduces interior habitat and may lead to the extinction of those species which require interior habitat. Restorative projects can increase the effective size of a habitat by simply adding area or by planting habitat corridors that link and fill in the gap between two isolated fragments. Reversing the effects of fragmentation and increasing habitat connectivity are the central goals of restoration ecology.

Ecosystem Function

Ecosystem function describes the foundational processes of natural systems, including nutrient cycles and energy fluxes. These processes are the most basic and essential components of ecosystems. An understanding of the full complexity and intricacies of these cycles is necessary to address any ecological processes that may be degraded. A functional ecosystem, that is completely self-perpetuating (no management required), is the ultimate goal of restorative efforts. Since, these ecosystem functions are emergent properties of the system as a whole, monitoring and management are crucial for the long-term stability of an ecosystem.

Evolving Concepts

Restoration ecology, because of its highly physical nature, is an ideal testing ground for an emerging community's ecological principles (Bradshaw 1987). Likewise, there are emerging concepts of inventing new and successful restoration technologies, performance standards, time frames, local genetics, and society's relationship to restoration ecology, and new ethical and religious possibilities, as future topics of discussion and debate.

Assembly

Community assembly "is a framework that can unify virtually all of (community) ecology under a single conceptual umbrella".Community assembly theory attempts to explain the existence of environmentally similar sites with differing assemblages of species. It assumes that species have similar niche requirements, so that community formation is a product of random fluctuations from a common species pool. Essentially, if all species are fairly ecologically equivalent then random variation in colonization, migration and extinction rates between species, drive differences in species composition between sites with comparable environmental conditions.

Stable States

Alternative stable states are discrete species compositional possibilities that may exist within a community. According to assembly theory, differences in species colonization, interspecific interactions and community establishment may result in distinct community species equilibria. A community has numerous possible compositional equilibria that are dependent on the initial assembly. That is, random fluctuations lead to a particular initial community assembly, which affects successional trajectories and the eventual species composition equilibrium.

Multiple stable states is a specific theoretical concept, where all species have equal access to a community (i.e., equal dispersal potential) and differences between communities arise simply because of the timing of each species' colonization.

These concepts are central to restoration ecology; restoring a community not only involves manipulating the timing and structure of the initial species composition, but also working towards a single desired stable state. In fact, a degraded ecosystem may be viewed as an alternative stable state under the altered environmental conditions.

Ontogeny

The ecology of ontogeny is the study of how ecological relationships change over the lifetime of an individual. Organisms require different environmental conditions during different stages of their life-cycle. For immobile organisms (e.g. plants), the conditions necessary for germination and establishment may be different from those of the adult stage. As an ecosystem is altered by anthropogenic processes the range of environmental variables may also be altered. A degraded ecosystem may not include the environmental conditions necessary for a particular stage of an organism's development. If a self-sustaining, functional ecosystem must contain environmental conditions for the perpetual reproduction of its species, restorative efforts must address the needs of organisms throughout their development.

Application of Theory

Restoration is defined as the application of ecological theory to ecological restoration. However, for many reasons, this can be a challenging prospect. Here are a few examples of theory informing practice.

Soil Heterogeneity Effects on Community Heterogeneity

Spatial heterogeneity of resources can influence plant community composition, diversity and assembly trajectory. Baer et al. (2005) manipulated soil resource heterogeneity in a tallgrass prairie restoration project. They found increasing resource heterogeneity, which on its own was insufficient to insure species diversity in situations where one species may dominate across the range of resource levels. Their findings were consistent with the theory regarding the role of ecological filters on community assembly. The establishment of a single species, best adapted to the physical and biological conditions can play an inordinately important role in determining the community structure.

Invasion, Competitive Dominance and Resource use

"The dynamics of invasive species may depend on their abilities to compete for resources and exploit disturbances relative to the abilities of native species". Seabloom et al. (2003) tested this concept and its implications in a California grassland restoration context. They found that the native grass species were able to successfully compete with invasive exotics, therefore, the possibility exists of restoring an original native grassland ecosystem.

Successional Trajectories

Progress along a desired successional pathway may be difficult if multiple stable states exist. Looking over 40 years of wetland restoration data Klotzi and Gootjans (2001) argue that unexpected and undesired vegetation assemblies "may indicate that environmental conditions are not suitable for target communities". Succession may move in unpredicted directions, but constricting environmental conditions within a narrow range may rein in the possible successional trajectories and increase the likelihood of a desired outcome.

Natural Capital Committee's recommendation for a 25-year plan

The UK Natural Capital Committee (NCC) made a recommendation in its second State of Natural Capital report published in March 2014 that in order to meet the Government's goal of being the first generation to leave the environment in a better state than it was inherited, a long-term 25-year plan was needed to maintain and improve England's natural capital. The UK Government has not yet responded to this recommendation.

The Secretary of State for the UK's Department for Environment, Food and Rural Affairs, Owen Paterson, described his ambition for the natural environment and how the work of the Committee fits into this at an NCC event in November 2012: "I do not, however, just want to maintain our natural assets; I want to improve them. I want us to derive the greatest possible benefit from them, while ensuring that they are available for generations to come. This is what the NCC's innovative work is geared towards'.

Ecosystem Restoration

Ecosystem restoration for the superb parrot on an abandoned railway line in Australia

Forest restoration in action at the Buffelsdraai Landfill Site Community Reforestation Project in South Africa

According to the Society for Ecological Restoration, ecosystem restoration is the return of a damaged ecological system to a stable, healthy, and sustainable state, often together with associated ecosystem services.

Rationale

There are many reasons to restore ecosystems. Some include:

- Restoring natural capital such as drinkable water or wildlife populations.
- Mitigating climate change (e.g. through carbon sequestration)
- Helping threatened or endangered species
- Aesthetic reasons (Harris et al. 2006, Macdonald et al. 2002)
- Moral reasons: we have degraded, and in some cases destroyed, many ecosystems so it falls on us to 'fix' them.

There are considerable differences of opinion in how to set restoration goals and how to definc their success. Some urge active restoration (e.g. eradicating invasive animals to allow the native ones to survive) and others who believe that protected areas should have the bare minimum of human interference. Ecosystem restoration has generated controversy, with skeptics who doubt that the benefits justify the economic investment or who point to failed restoration projects and question the feasibility of restoration altogether. It can be difficult to set restoration goals, in part because, as Anthony Bradshaw claims, "ecosystems are not static, but in a state of dynamic equilibrium.... [with restoration] we aim [for a] moving target.'

Even though an ecosystem may not be returned to its original state, the functions of the ecosystem (especially ones that provide services to us) may be more valuable than its current configuration (Bradshaw 1987). One reason to consider ecosystem restoration is to mitigate climate change through activities such as afforestation. Afforestation involves replanting forests, which remove carbon dioxide from the air. Carbon dioxide is a leading cause of global warming (Speth, 2005) and capturing it would help alleviate climate change. Another example of a common driver of restoration projects in the United States is the legal framework of the Clean Water Act, which often requires mitigation for damage inflicted on aquatic systems by development or other activities.

Challenges in Restoration

Some view ecosystem restoration as impractical, in part because it sometimes fails. Hilderbrand et al. point out that many times uncertainty (about ecosystem functions, species relationships, and such) is not addressed, and that the time-scales set out for 'complete' restoration are unreasonably short. In other instances an ecosystem may be so degraded that abandonment (allowing an injured ecosystem to recover on its own) may be the wisest option (Holl, 2006). Local communities sometimes object to restorations that include the introduction of large predators or plants that require disturbance regimes such as regular fires (MacDonald et al. 2002). High economic costs can also be perceived as a negative impact of the restoration process. Public opinion is very important in the feasibility of a restoration; if the public believes that the costs of restoration outweigh the benefits they will not support it (MacDonald et al. 2002). In these cases people might be ready to leave the ecosystem to recover on its own, which can sometimes occur relatively quickly (Holl, 2006).

Many failures have occurred in past restoration projects, many times because clear goals were not set out as the aim of the restoration. This may be because, as Peter Alpert says, "people may not [always] know how to manage natural systems effectively'. Also many assumptions are made about myths of restoration such as the carbon copy, where a restoration plan, which worked in one area, is applied to another with the same results expected, but not realized (Hilderbrand et al. 2005).

REFERENCES

1. Elliott, S., D. Blakesley and K. Hardwick, in press. Restoring Tropical Forests: a Practical Guide. Kew Publications, London
2. Sgró, C.M., A. J. Lowe and A. A. Hoffmann, 2011. Building evolutionary resilience for conserving biodiversity under climate change. Evolutionary Applications 4 (2): 326-337
3. Lamb, David (2011). Regreening the Bare Hills. Springer. p. 547. ISBN 978-90-481-9870-2.
4. Stanturf, John A. (2005). What is forest restoration?. *Restoration of boreal and temperate forests. Boca Raton: CRC Press. pp. 3–11.*
5. Nations, Food and Agriculture Organization of the United. "Assisted natural regeneration of forests'. www.fao.org. Retrieved 2016-02-24.
6. Martínez de Aragón, Juan; Fischer, Christine; Bonet, Jose-Antonio; Olivera, Antoni; Oliach, Daniel; Colinas, Carlos (2012). Economically profitable post fire restoration with black truffle (Tuber melanosporum) producing plantations'. New Forests. **43**: 615–630.
7. Mansourian, S., D. Vallauri, and N. Dudley (eds.) (in co-operation with WWF International), 2005. Forest Restoration in Landscapes: Beyond Planting Trees. Springer, New York
8. Reitbergen-McCraken, J., S. Maginnis A. Sarre, 2007. The Forest Landscape Restoration Handbook. Earthscan, London, 175 pp.
9. Shono, K., E. A. Cadaweng and P. B. Durst, 2007. Application of Assisted Natural Regeneration to Restore Degraded Tropical Forestlands. Restoration Ecology, 15(4): 620–626.
10. Elliott S, Navakitbumrung P, Kuarak C, Zankum S, Anusarnsunthorn V, Blakesley D, 2003. Selecting framework tree species for restoring seasonally dry tropical forests in northern Thailand based on field performance. For Ecol Manage 184:177-191
11. Goosem, S. and N. I. J. Tucker, 1995. Repairing the Rainforest. Wet Tropics Management Authority, Cairns, Australia. Pp 72. http://www.wettropics.gov.au/media/med_landholders.html

12. Regos, Adrian; D'Amen, M; Titeux, Nicolas; Herrando, Sergi; Guisan, A; Brotons, Lluis (2016). [10.1111/ddi.12375 "Predicting the future effectiveness of protected areas for bird conservation in Mediterranean ecosystems under climate change and novel fire regime scenarios'] Check |url= value (help). Diversity and Distributions. **22** *(1): 83–96. doi:10.1111/ddi.12375.*

13. Martín-Alcón, Santiago; Coll, Lluis (2016). "Unraveling the relative importance of factors driving post-fire regeneration trajectories in non-serotinous Pinus nigra forests' (PDF). Forest Ecology and Management. **361**: 13–22. doi:10.1016/j.foreco.2015.11.006.

14. City of Ashland, Oregon - Ashland Forest Resiliency Project - What is AFR?'. www.ashland.or.us. Retrieved 2016-02-05.

15. Ashland Forest Resilency Project - Lomakatsi Restoration Project'. Lomakatsi Restoration Project. Retrieved 2016-02-05.

16. Toman, E., Stidham, M., Shindler, B., McCaffrey, S. 2011. Reducing fuels in the wildland-urban interface: community perceptions of agency fuel treatments. Intl. Journal of Wildland Fire 20 (3):340-349

17. Hagmann, R.K., Franklin, J.F., Johnson, N.K. 2013. Historical structure and composition of ponderosa pine and mixed-conifer forests in south-central Oregon. Forest Ecology and Management 304: 492-504

18. Franklin, J.F., and Johnson, K.N. 2012. A restoration framework for federal forests in the Pacific Northwest. Journal of Forestry 110 (8): 429-439

19. City of Ashland, Oregon - Ashland Forest Resiliency Project - AFR Work Update. www.ashland.or.us. Retrieved 2016-02-05.

20. Reitbergen-McCraken, J., S. Maginnis A. Sarre, 2007. The Forest Landscape Restoration Handbook. Earthscan, London, 175 pp

21. Lamb, D., 2011. Regreening the Bare Hills. Springer 547pp.

22. Mansourian, S., D. Vallauri, and N. , Dudley (eds.) (in co-operation with WWF International), Forest Restoration in Landscapes: Beyond Planting Trees. Springer, New York.

23. Elliott, S., D. Blakesley and K. Hardwick, in press. Restoring Tropical Forests: a Practical Guide. Kew Publications, London

24. The Global Partnership on Forest and Landscape Restoration'. The Global Partnership on Forest and Landscape Restoration. IUCN.

25. Food and Agriculture Organization of the United Nations. Forest and Landscape Restoration Mechanism. Food and Agriculture Organization of the United Nations.

26. Global Partnership on Forest Landscape Restoration. The Bonn Challenge. The Bonn Challenge. International Union for Conservation of Nature.

27. Convention on Biological Diversity. Aichi Biodiversity Targets (PDF). Convention on Biological Diversity.

28. Sustainable Financing for Forest and Landscape Restoration. The Role of Public Policy Makers (PDF). Food and Agricultural Organization of the United Nations and the Global Mechanism of the UNCCD. June 2015.

29. Sustainable Financing for Forest and Landscape Restoration - Opportunities, challenges and the way forward (PDF). Food and Agriculture Organization of the United Nations and the Global Mechanism of the UNCCD. June 2015. ISBN 978-92-5-108992-7.

CHAPTER 8

Sustainable Forest Management

Chapter Contents

- **Introduction**
- **Scope**
- **Opportunities for Forest Restoration**
- **Natural Regeneration**
- **Post-Fire Regeneration**
- **Forest Restoration Projects**
- **Forest Landscape Restoratio**
- **Secondary Forest**
- **Restoration Ecology**

Sustainable forestry is a forestry practice which is designed to be sustainable in the long term, meaning that it can be continued into the foreseeable future without running out of resources or threatening the environment. This approach to forestry and forest management is increasingly popular in many regions of the world as consumer concern about the environment grows, driving companies which use environmental resources to think about environmental problems when developing business practices. Many forestry schools offer training in sustainable forestry for students who are interested in this practice.

INTRODUCTION

Sustainable forest management (SFM) is the management of forests according to the principles of sustainable development. Sustainable forest management has to keep the balance between three main pillars: ecological, economic and socio-cultural. Successfully achieving sustainable forest management will provide integrated benefits to all, ranging from safeguarding local livelihoods to protecting the biodiversity and ecosystems provided by forests, reducing rural poverty and mitigating some of the effects of climate change.

The "Forest Principles" adopted at The United Nations Conference on Environment and Development (UNCED) in Rio de Janeiro in 1992 captured the general international understanding of sustainable forest management at that time. A number of sets of criteria and indicators have since been developed to evaluate the achievement of SFM at the global, regional, country and management unit level. These were all attempts to codify and provide for independent assessment of the degree to which the broader objectives of sustainable forest management are being achieved in practice. In 2007, the United Nations General Assembly adopted the Non-Legally Binding Instrument on All Types of Forests. The instrument was the first of its kind, and reflected the strong international commitment to promote implementation of sustainable forest management through a new approach that brings all stakeholders together.

DEFINITION

A definition of SFM was developed by the Ministerial Conference on the Protection of Forests in Europe (FOREST EUROPE), and has since been adopted by the Food and Agriculture Organization (FAO). It defines sustainable forest management as:

The stewardship and use of forests and forest lands in a way, and at a rate, that maintains their biodiversity, productivity, regeneration capacity, vitality and their potential to fulfill, now and in the future, relevant ecological, economic and social functions, at local, national, and global levels, and that does not cause damage to other ecosystems.

In simpler terms, the concept can be described as the attainment of balance – balance between society's increasing demands for forest products and benefits, and the preservation of forest health and diversity. This balance is critical to the survival of forests, and to the prosperity of forest-dependent communities.

For forest managers, sustainably managing a particular forest tract means determining, in a tangible way, how to use it today to ensure similar benefits, health and productivity in the future. Forest managers must assess and integrate a wide array of sometimes conflicting factors – commercial and non-commercial values, environmental considerations, community needs,even global impact – to produce sound forest plans. In most cases, forest managers develop their forest plans in consultation with citizens, businesses, organizations and other interested parties in and around the forest tract being managed. The tools and visualization have been recently evolving for better management practices.

In 2014, at the request of Member States, the Food and Agriculture Organization of the United Nations developed and launched the Sustainable Forest Management Toolbox, an online collection of tools, case studies and other resources to support countries implementing sustainable forest management.

Because forests and societies are in constant flux, the desired outcome of sustainable forest management is not a fixed one. What constitutes a sustainably managed forest will change over time as values held by the public change.

CRITERIA AND INDICATORS

Criteria and indicators are tools which can be used to conceptualise, evaluate and implement sustainable forest management. Criteria define and characterize the essential elements, as well as a set of conditions or processes, by which sustainable forest management may be assessed. Periodically measured indicators reveal the direction of change with respect to each criterion.

Criteria and indicators of sustainable forest management are widely used and many countries produce national reports that assess their progress toward sustainable forest management. There are nine international and regional criteria and indicators initiatives, which collectively involve more than 150 countries. Three of the more advanced initiatives are those of the Working Group on Criteria and Indicators for the Conservation and Sustainable Management of Temperate and Boreal Forests (also called the Montreal Process), Forest Europe, and the International Tropical Timber Organization. Countries who are members of the same initiative usually agree to produce reports at the same time and using the same indicators. Within countries, at the management unit level, efforts have also been directed at developing local level criteria and indicators of sustainable forest management. The Center for International Forestry Research, the International Model Forest Network and researchers at the University of British Columbia have developed a number of tools and techniques to help forest-dependent communities develop their own local level criteria and indicators. Criteria and Indicators also form the basis of third-party forest certification programs such as the Canadian Standards Association's Sustainable Forest Management Standards and the Sustainable Forestry Initiative Standard.

There appears to be growing international consensus on the key elements of sustainable forest management. Seven common thematic areas of sustainable forest management have emerged based on the criteria of the nine ongoing regional and international criteria and indicators initiatives. The seven thematic areas are:

- Extent of forest resources
- Biological diversity
- Forest health and vitality
- Productive functions and forest resources
- Protective functions of forest resources
- Socio-economic functions
- Legal, policy and institutional framework.

This consensus on common thematic areas (or criteria) effectively provides a common, implicit definition of sustainable forest management. The seven thematic areas were acknowledged by the international forest community at the fourth session of the United Nations Forum on Forests and the 16th session of the Committee on Forestry. These thematic areas have since been enshrined in the Non-Legally Binding Instrument on All Types of Forests as a reference framework for sustainable forest management to help achieve the purpose of the instrument.

On January 5, 2012, the Montreal Process, Forest Europe, the International Tropical Timber Organization, and the Food and Agriculture Organization of the United Nations, acknowledging the seven thematic areas, endorsed a joint statement of collaboration to improve global forest related data collection and reporting and avoiding the proliferation of monitoring requirements and associated reporting burdens.

ECOSYSTEM APPROACH

The Ecosystem Approach has been prominent on the agenda of the Convention on Biological Diversity (CBD) since 1995 . The CBD definition of the Ecosystem Approach and a set of principles for its application were developed at an expert meeting in Malawi in 1995, known as the Malawi Principles. The definition, 12 principles and 5 points of "operational guidance" were adopted by the fifth Conference of Parties (COP5) in 2000. The CBD definition is as follows

The ecosystem approach is a strategy for the integrated management of land, water and living resources that promotes conservation and sustainable use in an equitable way. Application of the ecosystem approach will help to reach a balance of the three objectives of the Convention. An ecosystem approach is based on the application of appropriate scientific methodologies focused on levels of biological organization, which encompasses the essential structures, processes, functions and interactions among organisms and their environment. It recognizes that humans, with their cultural diversity, are an integral component of many ecosystems.

Sustainable forest management was recognized by parties to the Convention on Biological Diversity in 2004 (Decision VII/11 of COP7) to be a concrete means of applying the Ecosystem Approach to forest ecosystems. The two concepts, sustainable forest management and the ecosystem approach, aim at promoting conservation and management practices which are environmentally, socially and economically sustainable, and which generate and maintain benefits for both present and future generations. In Europe, the MCPFE and the Council for the Pan-European Biological and Landscape Diversity Strategy (PEBLDS) jointly recognized sustainable forest management to be consistent with the Ecosystem Approach in 2006.

INDEPENDENT CERTIFICATION

Growing environmental awareness and consumer demand for more socially responsible businesses helped third-party forest certification emerge in the 1990s as a credible tool for communicating the environmental and social performance of forest operations.

There are many potential users of certification, including: forest managers, scientists, policy makers, investors, environmental advocates, business consumers of wood and paper, and individuals.

With third-party forest certification, an independent organization develops standards of good forest management, and independent auditors issue certificates to forest operations that comply with those standards. Forest certification verifies that forests are well-managed—as defined by a particular standard—and chain-of-custody certification tracks wood and paper products from the certified forest through processing to the point of sale.

This rise of certification led to the emergence of several different systems throughout the world. As a result, there is no single accepted forest management standard worldwide, and each system takes a somewhat different approach in defining standards for sustainable forest management.

In its 2009–2010 Forest Products Annual Market Review United Nations Economic Commission for Europe/Food and Agriculture Organization stated: Over the years, many of the issues that previously divided the (certification) systems have become much less distinct. The largest certification systems now generally have the same structural programmatic requirements.

Third-party forest certification is an important tool for those seeking to ensure that the paper and wood products they purchase and use come from forests that are well-managed and legally harvested. Incorporating third-party certification into forest product procurement practices can be a centerpiece for comprehensive wood and paper policies that include factors such as the protection of sensitive forest values, thoughtful material selection and efficient use of products.

The ecolabel of the Forest Stewardship Council for certified sustainable forest product.

There are more than fifty certification standards worldwide, addressing the diversity of forest types and tenures. Globally, the two largest umbrella certification programs are:

- Programme for the Endorsement of Forest Certification (PEFC)
- Forest Stewardship Council (FSC)

The area of forest certified worldwide is growing slowly. PEFC is the world's largest forest certification system, with more than two-thirds of the total global certified area certified to its Sustainability Benchmarks.

In North America, there are three certification standards endorsed by PEFC – the Sustainable Forestry Initiative, the Canadian Standards Association's Sustainable Forest Management Standard, and the American Tree Farm System. FSC has five standards in North America – one in the United States and four in Canada.

While certification is intended as a tool to enhance forest management practices throughout the world, to date most certified forestry operations are located in Europe and North America. A significant barrier for many forest managers in developing countries is that they lack the capacity to undergo a certification audit and maintain operations to a certification standard.

FOREST GOVERNANCE

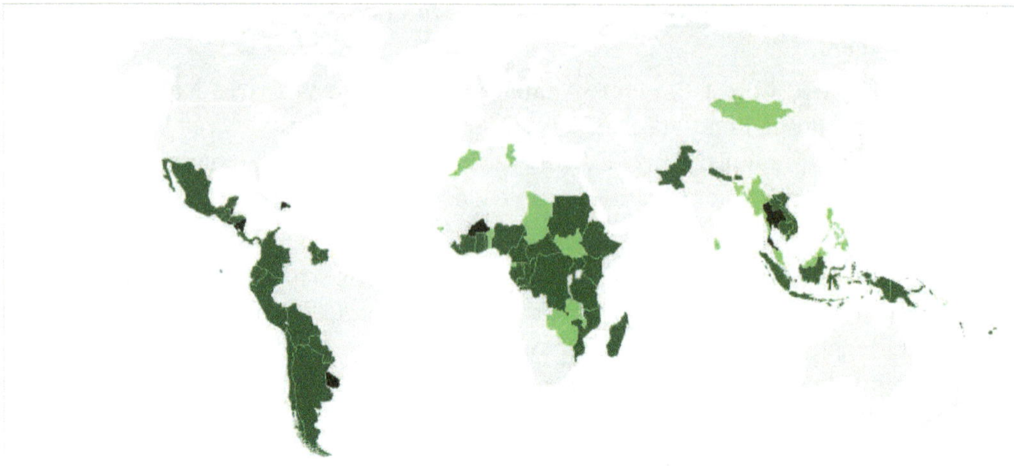

Countries participating in the UNREDD program and/or Forest Carbon Partnership Facility.

- UN-REDD participants
- Forest Carbon Partnership Facility participants
- participants in both

Although a majority of forests continue to be owned formally by government, the effectiveness of forest governance is increasingly independent of formal ownership. Since neo-liberal ideology in the 1980s and the emanation of the climate change challenges, evidence that the state is failing to effectively manage environmental resources has emerged. Under neo-liberal regimes in the developing countries, the role of the state has diminished and the market forces have increasingly taken over the dominant socio-economic role. Though the critiques of neo-liberal policies have maintained that market forces are not only inappropriate for sustaining the environment, but are in fact a major cause of environmental destruction. Hardin's tragedy of the common (1968) has shown that the people cannot be left to do as they wish with land or environmental resources. Thus, decentralization of management offers an alternative solution to forest governance.

The shifting of natural resource management responsibilities from central to state and local governments, where this is occurring, is usually a part of broader decentralization process. According to Rondinelli and Cheema (1983), there are four distinct decentralization options: these are: (i) Privatization – the transfer of authority from the central government to non-governmental sectors otherwise known as market-based service provision, (ii) Delegation – centrally nominated local authority, (iii) Devolution – transfer of power to locally acceptable authority and (iv) Deconcentration – the redistribution of authority from the central government to field delegations of the central government. The major key to effective decentralization is increased broad-based participation in local-public decision making. In 2000, the World Bank report reveals that local government knows the needs and desires of their constituents better than the national government, while at the same time, it is easier to hold local leaders accountable. From the study of West African tropical forest, it is argued that the downwardly accountable and/or representative authorities with meaningful discretional powers are the basic institutional element of decentralization that should lead to efficiency, development and equity. This collaborates with the World Bank report in 2000 which says that decentralization should improve resource allocation, efficiency, accountability and equity "by linking the cost and benefit of local services more closely".

Many reasons point to the advocacy of decentralization of forest. (i) Integrated rural development projects often fail because they are top-down projects that did not take local people's needs and desires into account. (ii) National government sometimes have legal authority over vast forest areas that they cannot control, thus, many protected area projects result in increased biodiversity loss and greater social conflict. Within the sphere of forest management, as state earlier, the most effective option of decentralization is "devolution"-the transfer of power to locally accountable authority. However, apprehension about local governments is not unfounded. They are often short of resources, may be staffed by people with low education and are sometimes captured by local elites who promote clientelist relation rather than democratic participation.Enters and Anderson (1999) point that the result of community-based projects intended to reverse the problems of past central approaches to conservation and development have also been discouraging.

Broadly speaking, the goal of forest conservation has historically not been met when, in contrast with land use changes; driven by demand for food, fuel and profit. It is necessary to recognize and advocate for better forest governance more strongly given the importance of forest in meeting basic human needs in the future and maintaining ecosystem and biodiversity as well as addressing climate change mitigation and adaptation goal. Such advocacy must be coupled with financial incentives for government of developing countries and greater governance role for local government, civil society, private sector and NGOs on behalf of the "communities".

National Forest Funds

The development of National Forest Funds is one way to address the issue of financing sustainable forest management.National forest funds (NFFs) are dedicated financing mechanisms managed by public institutions designed to support the conservation and sustainable use of forest resources. As of 2014, there are 70 NFFs operating globally.

EVEN-AGED TIMBER MANAGEMENT

Even-aged timber management is a group of forest management practices employed to achieve a nearly coeval cohort group of forest trees. The practice of even-aged management is often pursued to minimize costs to loggers. In some cases, the practices of even aged timber management are frequently implicated in biodiversity loss and other ecological damage.Even-aged timber management can also be beneficial to restoring natural native species succession.

Process

The first step in even aged timber management is to select a suitable stand for harvest. Trees must be of merchantable size, a desirable species, and in an area accessible to harvesting equipment. Once selected, the stand is harvested (usually using feller-bunchers, skidders, and processors). Merchantable trees (trees with boles large enough to be sold to a mill) are harvested and processed whereas unmerchantable trees (trees that are too small or of an undesirable species) are either crushed by machinery or cut to make equipment movement easier. This process is called clearcutting, and the end result is a cutblock. At this stage, the forestry company is required to replant the same proportion of species that were there previous to harvest. Usually, silviculture techniques are employed to ensure that species survive and grow in the proper proportion. Shade-intolerant species such as trembling aspen (*Populus tremuloides*) and marsh reedgrass (*Calamagrostis canadensis*) will thrive in the open conditions of a clearcut, outcompeting shade-tolerant species such as white spruce (*Picea glauca*). Mechanical site preparation and herbicide treatments are often applied initially to allow for adequate shade-tolerant growth. This way, the shade-tolerants are given a head start and the shade-intolerants will catch up over time once silviculture treatments are stopped, resulting in an even-aged stand of trees with little vertical stratification.

Economic Implications

Even-aged forest management is the harvesting system of choice in many parts of the world because it is often considered to be the only method that is economically viable. Forestry operations have extremely high variable costs- per hour expenses for harvesting equipment and per kilometer expenses for log transportation compose a very large portion of the total cost to harvest a stand. Therefore, companies must obtain as much productivity as possible from their hourly equipment in order to be profitable. Uneven-aged management techniques where only certain trees are harvested, such as shelterwood systems, incur the same hourly costs while harvesting fewer trees per hour. This reality explains why in Canada, for example, 90% of forests harvested are done so by clear cut through an even-aged management system.

Wildlife Considerations

Ecological analysis indicates that even aged timber management can produce inferior outcomes for wildlife biodiversity and abundance. Some species thrive on uneven or natural forest tree distribution. For example, the wild turkey thrives when uneven heights and canopy variations exist and its numbers are diminished by even aged timber management.

Depending on the ecosystem, however, some species thrive under even aged timber management. Clear cutting closely mimics the results of the most prominent natural disturbance in the boreal forest- wildfires. Certain tree species of this forest type have evolved over time to regenerate favourably in open, well-lit conditions, while others are adapted to thrive in the shade they produce. Clear cutting allows shade intolerant species, such as aspen and jack pine, to regenerate where they could not in a mature forest.

Clear cutting areas for even-aged timber management can also benefit many wildlife species. The abundance of coarse woody debris added to the forest floor is beneficial to salamander and snake species, while the release of understory shrubs provides an increase of berry-foraging opportunities for bear and songbirds. In general, the immature, grassy forest regenerated after a clear cut fulfills specific habitat requirements for many species that cannot be found in mature forests. In the boreal forest, for example, wildfire has been suppressed by humans, resulting in many overmature stands. Even aged timber management restores the successional cycle and allows early successional species to recolonize immature habitat.

HEALTHY FORESTS INITIATIVE

The **Healthy Forests Initiative** (or **HFI**), officially the **Healthy Forests Restoration Act of 2003** (P.L. 108-148), is a law proposed by President George W. Bush. Its stated intent is to reduce the threat of destructive wildfires. The law seeks to accomplish this by allowing timber harvests on protected National Forest's land. The law streamlined the permitting process for timber harvests in National Forests by adding new categorical exclusions to the National Forest Service's list of categorical exclusions from the environmental impact assessment process.

Supporters of the law contend that this will reduce wildfire risk by thinning overstocked stands, clearing away vegetation and trees to create shaded fuel breaks, providing funding and guidance to reduce or eliminate hazardous fuels in National Forests, improving forest fire fighting, and researching new methods to halt destructive insects. To proponents, much of the basis for the law revolves around the overcrowding of forests due to the suppression of low intensity fires. The resulting buildup of ground fuels and trees is thought to have increased the size and severity of wildfires in the United States.

Detractors of the law contend that the bill opens previously protected forest areas to logging, often unnecessarily or under false pretense. Disagreement exists concerning the role of private logging companies in thinning stands and clearing fire-breaks. The HFI also requires that communities within the "wildland urban interface" create "community wildfire protection plans." Community wildfire protection plans designate areas adjacent to communities that should be thinned so that crown fires will not directly burn into communities.

Controversy

The Bush administration claims broad support for HFI, stating on the official website: "The Administration and a bipartisan majority in Congress supported the legislation and are joined by a variety of environmental conservation groups." This statement ignores the opposition to HFI by conservation groups such as the Sierra Club, the Natural Resources Defense Council, The Wilderness Society, and the John Muir Project. Supporters include the Society of American Foresters,local fire protection agencies, and a number of hunting and fishing advocacy groups.

In March 2006, it was reported in the news section of the ACS journal *Environmental Science & Technology* that timber interests created a front group called Project Protect to help pass the Healthy Forests legislation.

HFI Support

Proponents of HFI believe that high intensity fires in the Western United States have increased in intensity and size in recent years. This is largely a result of fire suppression, which has created a buildup in fuels and ladder fuels, as well as climate change. Supporters believe this act may be able to simplify the NEPA process for projects which are meant to reduce the susceptibility of National forests to forest fires; reducing the risk of conflagrations that endanger human lives, destroy valuable forests, and can reduce the quality of habitat for late seral wildlife species.

The act is specific, that projects under the act should focus on small tree removal, and maximize retention of large trees. Furthermore, supporters of the act point to the fact that projects under the act must, before going through NEPA, be approved by a collaborative group, which should include all interested and affected parties (including adjacent land owners, special use permit holders in the area, and environmental groups). The intent of NEPA is to allow the public to have input on projects, influence the topics of analysis for projects, and develop alternatives for projects. Collaborative groups may allow for more input and influence on projects from the public than the normal NEPA process, despite the fact that projects under the act go through an expedited NEPA process.

The cost of fighting major wildfires caused by years of neglect have become entirely too expensive for USFS budgetary expenses. American Insurance companies have raised the issue of structural destruction caused by super forest fires is exceeding 2 billion dollars a year. From a historical standpoint looking to the Sierra Nevada Mountains prior to the inception of the Forest Service the forests were burned every year by the ranchers who ran livestock in these regions. Consequently the amount of fuel left on the forest floor was minor, resulting in low intensity fires. The fires were typically lit during the first storms of the winter and crept along the forest floor. Low branches and fallen trees were consumed in the low intensity larger logs tended to burn beneath the snow during the winter.

Positive advantages of this practical method of forest management was that the ash left behind fertilized the forest trees and plants and bark beetles which are ravishing so much of the Western Forests were killed before boring into trees. Pinecones were also liberated by the heat to release their seeds into the soil. In the spring natural grasses provided extensive grazing opportunities for livestock and wildlife inhabiting the forest regions and also provided critical anti-erosion during heavy rains. To return the forest to such pristine grandness will require extensive planning to remove the more dense fuel loads and careful prescribed burns to liquidate remaining pockets of heavy fuels left on the forest floors by 70 years of neglect.

HFI Opposition

Opponents of the nicknamed "No Tree Left Behind" Act point out that logging companies will be allowed to unnecessarily cut large diameter trees under a false pretense, while neglecting the greater issue of ladder fuels (such as brush and small trees) and possibly leaving debris that would add to extremely volatile ground fuels. Furthermore, forest fires occur naturally and are critical to the long-term survival of many forests, since many trees will only grow once they detect that a fire has occurred since this gives them access to sunlight. Some opponents also criticize the blanket prescription of thinnings to forests where low intensity fires did not historically play a pivotal role.

Furthermore, some saplings only grow after forest fires have cleared away older and dead trees; if humans intervene by preventing natural forest fires from occurring for too long, the forest will eventually die and grow back very slowly.

Litigation

In 2004, the Sierra Club and Sierra Forest Legacy (formerly named Sierra Nevada Forest Protection Campaign) brought a lawsuit challenging one aspect of HFI. The National Environmental Policy Act requires preparation of an environmental impact statement (EIS) for agency actions. Under HFI, the Forest Service had promulgated a "categorical exclusion" that eliminated the EIS requirement for timber sales up to 1,000 acres (4.0 km^2) and prescribed burns up to 4,500 acres (18 km^2). On December 5, 2007, in *Sierra Club v. Bosworth*, the Ninth Circuit held that the Forest Service's promulgation of the categorical exclusion "was arbitrary and capricious".

REFERENCES

1. LEDS GP Agriculture, Forestry and Other Land Use Working Group factsheet" (PDF). Low Emission Development Strategies Global Partnership (LEDS GP). Retrieved 23 March 2016.
2. Ministerial Conference on the Protection of Forests in Europe". Mcpfe.org. Retrieved 30 November 2011.
3. Evans, K., De Jong, W., and Cronkleton, P. (2008) "Future Scenarios as a Tool for Collaboration in Forest Communities". "S.A.P.I.EN.S." "'1'" (2)". Sapiens.revues.org. 1 October 2008. Retrieved 30 November 2011.
4. Mozgeris, G. (2008) "The continuous field view of representing forest geographically: from cartographic representation towards improved management planning". S.A.P.I.EN.S.1 (2)". Sapiens.revues.org. Retrieved 30 November 2011.
5. Sustainable Forest Management Toolbox" (PDF). Food and Agriculture Organization of the United Nations. Retrieved 24 June 2014.
6. Guidelines for Developing, Testing and Selecting Criteria and Indicators for Sustainable Forest Management Ravi Prabhu, Carol J. P. Colfer and Richard G. Dudley. 1999. CIFOR. The Criteria & Indicators Toolbox Series.

7. Criteria and Indicators for Sustainable Forest Management: A Compendium. Paper compiled by Froylán Castañeda, Christel Palmberg-Lerche and Petteri Vuorinen, May 2001. Forest Management Working Papers, Working Paper 5. Forest Resources Development Service, Forest Resources Division. FAO, Rome (unpublished).
8. Montreal Process Indicators
9. MCPFE indicators
10. ITTO
11. International Model Forest Network". Imfn.net. Retrieved 30 November 2011.
12. CIFOR Criteria and Indicators Toolbox Series". Cifor.cgiar.org. Retrieved 30 November 2011.
13. International Model Forest Network Criteria and IndicatorsArchived 23 October 2006 at the Wayback Machine.
14. SFM Indicator Knowledge Base
15. Canadian Standards Association". Csa-international.org. Retrieved 30 November 2011.
16. Sustainable Forestry Initiative Introduction Page 1"(PDF). Retrieved 30 November 2011.
17. United Nations Forum on Forests (2004)
18. Committee on Forestry (2003)
19. Malawi Principles Archived 23 December 2007 at the Wayback Machine.
20. MCPFE". MCPFE. Retrieved 30 November 2011.
21. Council". Strategyguide.org. Retrieved 30 November 2011.
22. Pan-European Biological and Landscape Diversity Strategy". Strategyguide.org. Retrieved 30 November 2011.
23. PEBLD Strategy" (PDF). Strategyguide.org. Retrieved 30 November 2011.
24. 2009–2010 Forest Products Annual Market Review Page 121" (PDF). Retrieved 30 November 2011.
25. Erin Malec (5 October 2011). "Forest Certification Resource Center". Metafore.org. Retrieved 30 November 2011.
26. UNECE/FAO Forest Products Annual Market Review
27. PEFC". PEFC. Retrieved 30 November 2011.
28. Sustainable Forest Initiative". Sfiprogram.org. Retrieved 30 November 2011.
29. Canadian Standards Association". Csa-international.org. Retrieved 30 November 2011.
30. American Tree Farm System". Treefarmsystem.org. 22 November 2011. Retrieved 30 November 2011.
31. Forest Stewardship Council (US)". Fscus.org. Retrieved 30 November 2011.

32. Forest Stewardship Council (Canada)". Fsccanada.org. Retrieved 30 November 2011.

33. Auer, M. (2012). 'Group forest certification for smallholders in Vietnam: An early test and future prospects'. Human ecology 40(1): 5–14.

34. Agrawal, A., Chhatre, A, and Hardin, R. (2008). 'Changing Governance of the World's Forest'. Science 320: 1460–1462

35. Lutz, E, and Caldecott, J. (1996). Decentralization and biodiversity: a world bank symposium. Washington: The World Bank.

36. Hague, M. (1999). 'The Fate of sustainable development under neo-liberal regime in developing countries', International political review 20(2): 197–218.

37. Stokke (1999). 'Sustainable development: A multi-faceted challenge' European journal of development 3(1): 8–31.

38. Margulis, S. . 'Decentralized environmental management', Annual World Bank Report.

39. Ribott, (1990). 'Accountable representation and power in Participatory and decentralized environmental management', Unasylva 50(4).

40. World Bank 1988

41. Lutz, E and Caldcott (1999) Decentralisation and biodiversity conservation, A World bank symposium. Washington: The World Bank

42. Enters and Anderson (1999) Rethinking the decentralisation and devolution of biodiversity conservation Unasylva 50(4)

43. Enters and Anderson (1999)

44. Rondinelli and Cheema (1981)Decentralization and development, sage publication, London

45. M, Larson "Natural resource and degradation in Nicaragua: Are local governments up to the job?

46. Larson (2002)

47. Rondinelli and Chaeema (1999)

48. 2012 STUDY ON FOREST FINANCING (PDF). Advisory Group on Finance Collaborative Partnership on Forests. June 2012.

49. Matta, Rao (2015). Towards effective national forest funds, FAO Forestry Paper 174 (PDF). Rome, Italy: Food and Agriculture Organization of the United Nations. ISBN 978-92-5-108706-0.

50. Matta, Rao (2015). Towards effective national forest funds, FAO Forestry Paper 174 (PDF). Rome, Italy: Food and Agriculture Organization of the United Nations. ISBN 978-92-5-108706-0.

Index